Tribals and Modern Agriculture

Tribals and Modern Agriculture

Dynamics of Fertiliser Use in Madhya Pradesh

D.C. Sah

RAWAT PUBLICATIONS

Jaipur • New Delhi • Bangalore • Mumbai • Hyderabad • Guwahati

ISBN 81-316-0305-9

The publication of this book was financially supported by ICSSR and the responsibility for the facts stated, opinions expressed, or conclusions reached, is entirely that of the author and that ICSSR accepts no responsibility for them.

ICSSR Consultant: Professor Amita Shah

Published by
Prem Rawat for **Rawat Publications**
Satyam Apts, Sector 3, Jawahar Nagar, Jaipur 302 004 (India)
Phone: 0141 265 1748/7006 Fax: 0141 265 1748
E-mail: info@rawatbooks.com
Website: www.rawatbooks.com

New Delhi Office
4858/24, Ansari Road, Daryaganj, New Delhi 110 002
Phone: 011 2326 3290

Also at Bangalore, Mumbai, Hyderabad and Guwahati

Typeset by Rawat Computers, Jaipur
Printed at Nice Printing Press, New Delhi

To

Gunvant M. Desai

who, in his no-nonsense ways, taught me

ignorance is not bliss.

Contents

Preface

Fertilisers in Indian agriculture became essential input once it became clear that area as a source of growth of agricultural production has exhausted. Synergetic effect of irrigation, seed and fertiliser became the strategy of increasing yield. Fertilisers over the years, thus, have become indispensable. During last fifty years, the fertiliser use has increased from 0.1 million tonnes to 17 million tonnes. By any standard, this is an impressive growth. The growth of fertiliser use has happened as more and more farmers adopted its use, diffused it to unfertilised area and started using fertilisers at higher rates. In the process, fertiliser use, which was confined to irrigated areas, slowly started moving in unirrigated areas as well. But still, the spread and rate of fertiliser use on unirrigated areas is a fourth of the use on irrigated areas. Western tribal belt of Madhya Pradesh is important not only for creating equilibrium of fertiliser uses across geographical locations in the state but also is the most likely source for the future growth. The present study is diagnostic in nature and focuses on economic forces affecting the cultivators' adoption and diffusion process of fertiliser use and analysis of dynamics of fertiliser use for three continuous years between 2001–02 and 2003–04. A number of processes could be responsible for lowering the growth in fertiliser use; the non-users could be only one of the inefficiencies. Equally important could be identification of the inefficiency in fertiliser use caused by infrequent users and inefficiency in fertiliser use caused by continuous users.

This monograph is an offshoot of a larger study on *Fertiliser Use in Indian Agriculture* being conducted at the MPISSR. We are grateful to the Indian Council of Social Science Research (ICSSR), New Delhi for financially supporting us for undertaking this research. Analysing growth in fertiliser use at micro level, the study tries to understand why farm after farm fertilisers in the remote tribal areas are being used so inefficiently. In the process, the study documents experiences of farmers who were unable to use fertilisers on all crops, were infrequent users and used fertilisers continuously at about 40 per cent of its potential. Understanding fertiliser use this way assumes significance, for it recognises that it is farm condition rather the farmer's condition that constrains fertiliser use.

Unlike irrigated Madhya Pradesh, supply push by the fertiliser distribution system is the main source of growth in fertiliser use in this area. Intensity of irrigation here is low and spread of HYVs is extremely poor. In fact, the available seed-fertiliser technology has bypassed this remote tribal area. In absence of a substantial growth in irrigation and limited scope of diffusion of HYVs, the supply side has become an important force in generating growth in fertiliser use in this area. The analysis reveals that at present the sample farmers are using about 28,000 kg of NPK. It is estimated that about 32,000 kg of additional NPK use can be expected in near future if extension system tries to minimise the inefficiency in fertiliser use. We find that about 80 per cent of total potential is remaining untapped owing to inefficient use by continuous users, 16 per cent remains untapped owing to non-users and 4 per cent owing to infrequent users who dropout of fertiliser use frequently. This expected addition to fertiliser use is about 49 per cent of the total potential for NPK use of the sample farmers. Thus, in western tribal belt of Madhya Pradesh, non-adoption of fertilisers, non-use of fertilisers on a crop and infrequent fertiliser use is not a serious problem. Bulk of the inefficiency is owing to fertiliser use at relatively low rates. We also find that improving the efficiency of fertiliser use of continuous users on cotton, maize, jowar and wheat is key to achieving over 80 per cent of untapped potential. The task of the extension system is now much easier.

A portion of the inefficiency is on account of farmers' perceptions that optimum use on a crop is actually lower than what is recommended by the extension agency, even under very difficult conditions of remote tribal areas. But a part of the inefficiency could also be owing to variations of the conditions that influence fertiliser use; the farm conditions. The findings also reveal that like adoption of fertilisers, its pattern of use too is governed by farm characteristics rather than farmer's characteristics. Improving (a) the soil-moisture of farm, (b) ways to apply fertilisers, and (c) access to institutional markets for input-output are far more important in converting non-adopters of fertilisers to adopters and changing a discontinuous use pattern towards a continuous use pattern. The need, therefore, is to calibrate the extension messages, access to credit on the one hand, and productivity gains through public investment on the other. Despite the fact that fertiliser prices for these farmers is a non-issue, a large number of farmers have chosen to raise price issue to voice their grievances against the neglect of the area especially with respect to agricultural infrastructure, level of available technology, low productivity and high cost and high risk associated with private investment in irrigation. Nonetheless, independent of fertiliser prices, a number of other and equally important issues are thrown up by our investigation. Forces like lack of irrigation facilities and consecutive crop failure, poor soil with low agricultural productivity have resulted in creating perceptions that low rates of fertiliser use is a better risk management device in uncertain yield-response environment.

The study owes a great deal to several people who have provided intellectual foundation to this research, those who have supported in conducting the research and those who have had an opportunity to go through the draft and provided critical comments. Gunvant M. Desai, A. Vaidyanathan, Pradaman Kumar, Vidya Sagar, Cristina David, N.V. Nampoothiri, Amita Shah, Prem Narayan, Suman Rustagi, S.K. Raheja, S.S. Acharya, Richa Singh, Ashok Gulati and R.P. Singh through their work have facilitated us and clarified many of our misconceptions. Special thanks are due to Gunvant M. Desai, Ashish Bhatt, Neeta Tapan and Amita Shah whose remarks and comments were valuable inputs for the study during its conception stage. Manu Gautam has critically commented on some of the chapters as they were developing. Comments by Amita Shah, B.R. Purohit and an anonymous referee have helped us in revising the earlier drafts. I have tried to incorporate their suggestions as much as possible but for interpretations, errors and omissions I only am responsible. Data collection in the remote tribal area is a challenge. But Anand Srivastava, Neeraj Jain, Khuman Singh, Antar Singh, Manoj Parmar, Om Prakash Rathod, Gurvendra Singh, Dhirendra Uarmalia, and Ashok Namdev have patiently helped us generate qualitative and quantitative data. A valuable help came in the form of Ashish Bhatt, who agreed to coordinate the data collection. Professor R.S. Gautam, President, Madhya Pradesh Institute of Social Science Research was extremely encouraging and helpful in arranging all kinds of support. I would also like to thank L.P. Gautam, Sunil Chandel, Yatindra Singh, Sandeep Joshi, Ashish Bhatt and Bhawar Sharma for providing logistics as well intellectual support. Sudeep Mishra and Puneet Gautam have provided word processing support and also saw that the monograph is brought out in this form. And finally yet importantly, our sample farmers have supported us in completing our data collection work. This monograph would not have taken shape without their support.

Chapter 1

Limits of Growth in Agriculture in Less Developed Areas

Dualistic models have offered increasing insights into the process of economic development in developing countries. They can be grouped into two classes. The first group[1] deals with an economy in which one of the two sectors, the agricultural sector is traditionally producing wage goods (foodgrains) and has unlimited supply of labour. The marginal productivity of labour is zero, sometimes even negative, implying that the labour in this sector is paid at a rate that is higher than its productivity. The non-agricultural sector is modern and draws labour from the reservoir of cheap labour supply, the agricultural sector. This transfer of labour from the traditional sector to modern sector does not result in decline in the production of the former. If wages in non-agricultural sector are equal to labour productivity, then the aggregate production increases with the transfer of labour between the sectors. This is the engine of growth in these models. An important omission in these classical models is that they have not given importance to technological changes in agriculture.[2]

Agricultural sector in the second group of models has received more importance than what it received in classical models. This new strategy of economic development emphasises land augmenting technological changes in agriculture and stresses on identifying the factors that determine the process of technological changes.[3] These views came into existence due to the growing recognition that different sectors of the economy are interdependent, and understanding the determinants of growth in agricultural output is essential if this sector has to play its role in economic development.

Insufficient foodgrain production and their large-scale imports[4] in late 1950s made it clear that agricultural sector has to perform better in order to develop the Indian economy.[5] Inelastic land supply on the one hand and access to modern seeds of some foodgrains on the other, resulted in extensive adoption of land augmenting technology in mid-1960s.[6] This strategy centres around modern agricultural inputs which have synergistic effect on the yields. In the process, the importance of chemical fertilisers grew enormously.[7]

For technologically dynamic agriculture, rapidly increasing fertiliser use cannot be overemphasised. Fertilisers have played an increasingly important role in the agricultural development of several developing countries.[8] In India, fertiliser use has helped in abating the increasing land constraints by improving the productivity.[9] This, however, does not call for complacence. Plan after plan, foodgrain production has remained below the targeted level. The failure to achieve the targeted growth of production was mainly because the additions to fertiliser use in different plans have remained below the targeted growth.[10]

The projected population and food requirement at the end of eleventh plan set by the Ninth Five-Year Plan reveals that foodgrain production needs to be about 337 million tonnes at the end 2011–12. This production has to be achieved by stepping up the fertiliser use to 45.48 million tonnes of plant nutrients.[11] At present, when fertiliser use in the country is about 16.1 million tonnes, assessing its growth as desired in

2000s needs clear understanding of factors which have effected it in past. Policy interventions needed for creating a favourable environment for fertiliser use to grow at a desirable pace calls for clear understanding of the factors which have constrained and augmented the past growth in fertiliser use. This diagnostic analysis becomes all the more relevant, for all locations do not contribute to this growth, especially under new price environment.

Fertiliser industry is facing an uncertain policy environment since 1992. Although fertiliser subsidy in the period has grown from Rs 44,000 million to Rs 1,33,000 million, the industry is riddled with problems like ill-health, stunted growth and highly unbalanced nutrient consumption. Ceiling on subsidy has become inevitable to keep the overall deficit within•reasonable limits. Urea pricing has pushed many efficient firms towards sickness. Moreover, domestic production of phosphatic and complex fertilisers has been at a disadvantage vis-à-vis their import. Despite this, the per hectare fertiliser use in Madhya Pradesh has increased from 0.81 million tonnes on NPK in 1990–91 to 1.2 million tonnes in 1999–2000. However, of late the consumption is stagnating and it is least likely that this growth of fertiliser use can be sustained for long without taking corrective steps at various levels.

A large proportion of growth in fertiliser use in Madhya Pradesh has taken place in irrigated regions where the potential for further growth is low. Tribal belt, a less endowed region, with low irrigation and uncertain weather conditions, has contributed to this growth process; the western tribal belt of Madhya Pradesh consumes as high as 60 kg of NPK per acre in comparison to about 95 kg in the heartland of Malwa region. Apparently, it seems that happenings in tribal belt constrain the dynamics of fertiliser use of the region to a considerable extent. A close scrutiny reveals the following peculiarities of the region with respect to fertiliser use. First, a large proportion of fertiliser is consumed during kharif, which is the main agricultural season of the region. Second, due to frequent crop failures the cropping pattern is quite diversified in this tribal belt. Third, a high proportion of fertiliser consumed in the region is in the form of nitrogen; phosphate use is prevalent on oilseeds and gram. Fourth, a large proportion of potential for fertiliser use in this area is still untapped. And lastly, this area not only does contribute in growth of fertiliser use but also is responsible for all the fluctuations and stagnation in fertiliser use even during normal weather years.

The above discussion clearly brings out the contours of future fertiliser growth in Madhya Pradesh. Tribal belt in Madhya Pradesh is important not only for creating equilibrium of fertiliser use across geographical locations in the state but it is also the likely source for the future growth. In other words, the future growth in fertiliser use shall be relatively low in highly irrigated areas of Madhya Pradesh and more in non-traditional areas where a relatively large portion of the potential for fertiliser use is still untapped. The following questions are important in this context. First, how to increase the growth of fertiliser use in these unirrigated areas? Second, how to increase the potential of fertiliser use in the area? Lastly, to understand the source of growth in fertiliser use in this tribal belt.

Fertiliser use is an outcome of two variables, proportion of area under different crops receiving fertilisers and the rate of fertiliser application. Since different crops respond differently to fertiliser use and also because of difference in the crop prices, a varying pattern of fertiliser use would emerge in different agro-climatic zones in the country. Fertiliser use would increase even in comparatively unfavourable price environment provided (1) the increase in potential for fertiliser use due to technological changes more than offsets the adverse effect of change in price environment; (2) the marginal returns to move from low level towards optimum level of fertiliser use are economically attractive to cultivators; and (3) the cultivators' expectations regarding the physical productivity of fertiliser use have improved.

The present study is diagnostic in nature and focuses on economic forces affecting the cultivators' adoption and diffusion process of fertiliser use in relatively stagnating tribal belt of western Madhya Pradesh. This study has attempted to analyse the dynamics of fertiliser use, after adopting it for three continuous period of time, 2001–02, 2002–03 and 2003–04. A number of processes could be responsible for lowering the growth in fertiliser use like non-use of fertilisers by farmers. But the non-use could be only one of the inefficiency. Equally important shall be to identify the inefficiency in fertiliser use caused by infrequent users and inefficiency in fertiliser use caused by continuous users. As non-price factors behind the process have been more important than price factors (Desai, 1986), the analysis needs to be taken up crop by crop rather than at the farm level. It should also be recognised that these forces behind the growth in fertiliser use in the tribal belt cannot be deciphered correctly without taking into cognisance the macro-policy environment in which fertiliser use is growing. This becomes all the more important, for both demand for and supply of fertilisers has remained under state control. The study, therefore, tries to locate the major analysis within the macro-policy environment of fertiliser use.

The study attempts to understand the following three processes:

> First, to understand the pattern of adoption of fertilisers in different crops in this tribal area and evaluate the time lag, after adoption, farmers did experience in diffusing fertilisers.
>
> Second, to understand the patterns in fertiliser use on different crops for non-users, infrequent fertiliser users as well as continuous users in the three years between 2001–02 and 2003–04.

Thirdly, to estimate and examine the quantum of most viable untapped potential for fertiliser use and factors explaining the patterns observed at the farm level.

The study is organised into eight chapters, the first one being the introduction. Chapter 2 gives an overview of the growth in fertiliser use in India and Madhya Pradesh and presents a review of the literature in order to understand how available empirical research has addressed issues relating to growth of fertiliser use. Two main objectives of this chapter are to highlight how the importance of fertiliser use has increased in agricultural production programmes and to present the exceptional features associated with past growth of fertiliser use. Explanation to these peculiarities is sought in the available literature on economics of fertiliser use. In Chapter 3, an analytical framework is developed to examine the questions emerging out of the review of literature. The specific questions addressed in this study are given next and are answered within this framework in subsequent chapters. Finally, the chapter presents the methods used in the analysis. Chapter 4 introduces the readers to the study area, develops an understanding relating to marketing and extension environment within which the fertiliser use has grown in the region and examines the pattern of fertiliser use among the sample cultivators.

Chapters 5, 6 and 7 present the analysis of fertiliser use at micro level. Chapter 5 first tries to understand the time dimension of adoption of fertilisers and then examines factors behind delay in adoption of fertiliser by some cultivators. Chapter 6 estimates how the NPK use patterns amongst sample farmers have changed between 2001–02 and 2003–04 and then examines the sources of growth in fertiliser uses on different crops. In Chapter 7, the micro-level understanding of previous chapters is extended to estimate and examine the untapped potential for fertiliser use that still exists among the sample farmers. The analysis provides understanding on the most viable proportion of untapped potential existing with the sample farmers. The chapter also analyses how the inefficiency in fertiliser use at the different levels can be minimised. It is argued that a portion of the inefficiency is because of farmers' perceptions that optimum use on a crop is actually lower than what is recommended by the extension agency. But, a large part of the inefficiency could be owing to variations of the conditions that influence fertiliser use; the farm conditions. Chapter 8 summarises the main findings and highlights the policy implications of these findings.

Notes

1. W. Arthur Lewis, "Economic Development with Unlimited Supplies of Labour" in A.N. Agrawal and S.P. Singh (eds.), *The Economics of Underdevelopment*, Oxford University Press, London, 1956, pp. 400–49; Guster Ranis and John Fei, "A Theory of Economic Development", *American Economic Review,* September 1961, pp. 533–65; Ragnar Nurkse, "Some International Aspects of Problems of Economic Development" in A.N. Agrawal and S.P. Singh (eds.), *The Economics of Underdevelopment,* op. cit., pp. 256–71. And also see John W. Mailer, "Models of Economic Growth and Land Augmenting Technological Changes in Foodgrain Production" in Nurul Islam (ed.), *Agricultural Policy in Developing Countries*, John Wiley & Sons, New York, 1974, pp. 20–27.
2. The neo-classical dualistic models do consider technological change in agriculture as a crucial factor in economic development but such changes are assumed as neutral. This, however, is in contrast to the realities where introduction of modern varieties has proved to be land augmenting technological changes. See D.W. Jangenson, "Development of a Dual Economy", *Economic Journal,* Vol. 71, June 1961, pp. 309–34; also see Mellor, op. cit., p. 23.
3. John W. Mellor and Uma Lele, "A Labour Supply Theory of Economic Development", Occasional Paper No. 43, Cornell University Press, 1971; Y. Hayami and V. Ruttan, *Agricultural*

Development in International Perspective, Johns Hopkins Press, Baltimore, 1971.

4. If the agricultural output barely keeps with population growth, it not only means lack of employment and income generation but also inadequate supply of wagegoods to non-agricultural sector, which adversely affects its performance. It is now widely recognised that if any one sub-sector of Indian economy has constrained the progress of the other, it is stagnant agriculture that has limited the performance of industrial sector. See Gunvant M. Desai, "Agricultural Sector and Growth Rate of NNP, 1951–52 to 1975–76", Working Paper No. 168, Indian Institute of Management, Ahmedabad, 1977.
5. P.R. Bramanand and C.N. Vakil, *Planning for an Expanding Economy*, Vora & Co., Bombay, 1960; and John W. Mellor, *New Economics of Growth,* Cornell University Press, Ithaca, 1976, pp. 2–6.
6. In this strategy the modern varieties of seeds (HYVs) and new cultivation practices become means of reducing land constraints. The process involves: (a) Land and water development, (b) Biochemical means of enriching soil-fertility and protection of plants from diseases, and (c) Designing and adapting new biological efficient crop varieties. The success of such a technological import, however, depends on capacity to develop and adapt new location-specific biotechnology by the technology importing country. See Y. Hayami and V. Ruttan, op. cit., pp. 50–53; V. Ruttan, "Technology Transfer, Institutional Transfer and Induced Technical and Institutional Change in Agricultural Development" in Lioyd B. Reyndds (ed.), *Agricultural Development Theory,* Yale University Press, London, 1975, pp. 167–69.
7. Ministry of Food and Agriculture, *Reorientation of Programmes of Agricultural Production,* GoI, New Delhi, 1965, pp. 5–12; also see Gunvant M. Desai, "Fertilisers in India's Agricultural Development", in C.H. Shah and C.N. Vakil (eds.), *Agricultural Development in India: Policy and Problems*, Orient Longman, Bombay, 1978, pp. 380–81.
8. It is estimated that in developing economies of Africa, Latin America, Near East and Far East, about 30 per cent of total grain production increase between 1948–52 and 1972–73 was achieved due to fertiliser use. In India, increase in fertiliser use resulted in

about 53 per cent of increase in foodgrain production between 1960–61 and 1970–71. The increasing importance of fertiliser use could be judged from the fact that during 1950s the contribution of fertilisers to food-grain production growth was only 10 per cent. See M.S. Mudhar and P.P. Andersen, "Fertiliser Policy Issues and Implications in Developing Countries", paper presented in 'FAI-IFDC' Fertiliser Seminar on Trends in Consumption and Production organised by Fertiliser Association of India, December 1–3, 1977; John W. Mellor, *New Economics of Growth*, op. cit., p. 49.

9. The annual growth in foodgrain production between 1964–65 and 1970–71 was about 18 per cent higher than what it was between 1949–50 and 1960–61. This jump in production after the introduction of HYVs has been achieved due to increase in productivity. John W. Mellor, *New Economics of Growth,* op. cit., pp. 48–49.
10. This is not to underestimate the contribution of other factors in constraining the agricultural growth. See A. Vaidyanathan, "Performance and Prospects of Crop Production in India", *Economic and Political Weekly*, Vol. 12 (33 and 34), Special Number 1977, pp. 1359–61.
11. Planning Commission, *Ninth Five-Year Plan*, Government of India.

Chapter 2

Growth of Fertiliser Use in Indian Agriculture

This chapter serves as a setting to the problem investigated in the present study. It presents how the importance of fertilisers in Indian agriculture has grown, and how fertiliser use has increased. This overview of growth raises some important questions. At the end of the chapter, answers to these questions are sought for in the available literature. In developing countries, fertiliser use exists because of the following forces: (1) the state's recognition of fertilisers in various agricultural production programmes; (2) development of institutions for its procurement, distribution, and use; and (3) the cultivators' expectations regarding the profitability of its use. In India, the need for fertiliser use was recognised in late nineteenth century but its use on non-plantation agriculture started only after the Grow More Food (GMF) campaign in early 1940s. After independence, fertiliser use has grown rapidly alongwith growing importance of this input in production process.

Recognition of Fertilisers

Plant nutrient deficiencies in Indian soils were recognised as far back as 1890s.[1] But till early 1940s the use of chemical fertilisers was propagated on non-plantation agriculture with caution.[2] With the introduction of Grow More Food campaign in 1942, fertiliser use received recognition as an important input for rapidly increasing foodgrain production. The GMF campaign worked as an administrative umbrella for several schemes to increase production of foodgrains. One of these was to propagate and increase the use of manure and fertilisers. Although, because of the supply constraints imposed by the war conditions, fertiliser use did not increase significantly in the early years of the GMF campaign, the government of independent India inherited both the recognition of fertilisers' importance in raising food production and also the institutional apparatus to push its use in the non-plantation agriculture.

After independence, the First Plan gave prime importance to manures. The plan, however, stressed that in order to provide nourishment to poor soil, increasing use of chemical fertilisers need not wait for full mobilisation of manure.[3] Moreover, the First Plan designed experiments to evaluate fertiliser response under varying soils in selected community project areas. In the Second Plan, the 'yardstick' approach of increasing foodgrain production was adopted. As a result, the importance to fertilisers was stressed in increasing foodgrain production. The plan envisaged that increased fertiliser use would account for about 25 per cent of additional foodgrain output.[4] During subsequent plans, the importance of fertilisers increased substantially. In the Third Plan, the importance of fertilisers grew out of the fact that expansion of land as a source of output growth was getting exhausted and other yield increasing inputs required considerable time for development. Consequently, in a supply-oriented Intensive Agricultural District Programme (IADP), fertiliser became the most important input in the package of inputs.[5] Thus, even before the introduction of HYVs, fertiliser was an important component in various programmes.

With the introduction of HYV programme in the mid-1960s, the importance of fertilisers increased further.[6] But, till the end of the Fifth Plan the approach was to encourage fertiliser use on HYVs under irrigated conditions where response of crops to fertiliser use was comparatively high.[7] In the Sixth Plan, the emphasis shifted towards extensive spread of fertilisers at low rates of application to cover larger farming area.[8] Although, chemical fertilisers are still important, as a strategy for agricultural development, of late a judicious nutrient management, whereby organic, inorganic and wormy fertilisers have gained importance. The growing importance of fertilisers at the level of planning could be further highlighted by considering their targets fixed in different Plans (Table 2.1). Although the actual fertiliser use remained much lower than the targeted use, the strategy

Table 2.1

Target of Fertiliser Use and Achievement in Different Plans

Plan		*'000 tonnes of NPK*		
		Target	*Achievement*	*Percentage achievement*
I	1950–51 to 1955–56*	117	131	112
II	1956–57 to 1960–61*	535	294	55
III	1961–62 to 1965–66	1,630	785	48
Plan holidays		2,800	1,761	63
IV	1969–70 to 1973–74	5,500	2,839	52
V	1974–75 to 1978–79	4,800	5,117	107
VI	1980–81 to 1984–85	9,650	8,211	85
VII	1985–86 to 1989–90	13,500	11,568	85
VIII	1992–93 to 1996–97	18,300	14,308	78
IX	1997–98 to 2001–02	20,000	17,360	87
X	2002–03 to 2006–07	30,000	NA	NA

*Only nitrogen and phosphate.

Sources: Sah (1982) and FAI, Fertiliser Statistics, Various Issues.

indicates the supply-oriented approach in fertiliser use, which calls for adequate measures for enlarged fertiliser availability and their adequate allocation among different states. The revised and low targets of fertiliser use in the Fifth Plan[9] highlight the recognition of importance of cultivators' demand. This recognition was also reflected in subsequent Plans and is an indication of the fact that untapped potential of fertiliser use has been in difficult geographical regions. But a realistic assessment of effective demand is consistent with subsequent deregulation of the fertiliser industry that took place after 1992.

To understand the past growth in fertiliser use it is necessary not only to note the above developments with respect to the recognition of fertilisers' importance in Indian agriculture but also with respect to total availability of fertilisers, their efficient distribution, and cultivators' demand for this input. The scope of this analysis, however, is confined to issues related to cultivators' demand for fertilisers.

Growth in Fertiliser Use

The past growth of fertiliser use in independent India has been phenomenal (Table 2.2). However, fertiliser use in non-plantation agriculture in early 1940s, as mentioned earlier, was tied up with recognition of fertiliser's importance under GMF campaign and creation of Central Fertiliser Pool (CFP) for supply arrangements. About 80 per cent fertilisers available for consumption that time were imported. CFP was established to import fertilisers, procure the entire quantity of domestic production, and to ensure equitable distribution of available fertilisers throughout the country at reasonable prices.

After independence, the government helped actively in popularising the fertiliser use. Firstly, the dependence on international markets for fertiliser supply was reduced by

Table 2.2
Fertiliser Consumption in India, 1952–53 to 2004–05

Year	Consumption ('000 tonnes)				Annual change (%age)			
	N	P	K	Total	N	P	K	NPK
1952–53	58	5	3	66	—	—	—	—
1953–54	89	8	8	105	53	60	167	59
1954–55	95	15	11	121	7	88	38	15
1955–56	107	13	10	131	13	–13	–9	8
1956–57	123	16	15	154	15	23	50	18
1957–58	149	22	13	184	21	38	–13	19
1958–59	172	30	22	224	15	36	69	22
1959–60	229	54	21	305	33	80	–5	36
1960–61	212	53	29	294	–7	–2	38	–4
1961–62	250	61	28	338	18	15	–3	15
1962–63	338	83	36	452	33	36	29	34
1963–64	377	117	51	544	13	41	42	20
1964–65	555	149	69	773	47	27	35	42
1965–66	575	133	77	785	4	–11	12	2
1966–67	738	249	114	1,101	28	87	48	40
1967–68	1,035	335	170	1,539	40	35	49	40
1968–69	1,209	382	170	1,761	17	14	0	14
1969–70	1,356	416	210	1,982	12	9	24	13
1970–71	1,479	541	236	2,256	9	30	12	14
1971–72	1,798	558	300	2,656	22	3	27	18
1972–73	1,839	581	348	2,768	2	4	16	4
1973–74	1,829	650	360	2,839	–1	12	4	3
1974–75	1,766	472	336	2,573	–3	–27	–7	–9
1975–76	2,149	467	278	2,894	22	–1	–17	12
1976–77	2,457	635	319	3,411	14	36	15	18
1977–78	2,913	867	506	4,286	19	37	59	26
1978–79	3,420	1,107	592	5,118	17	28	17	18
1979–80	3,499	1,150	606	5,256	2	4	3	3
1980–81	3,678	1,214	624	5,516	5	6	3	5

Cont'd...

Cont'd...

1981–82	4,069	1,322	676	6,067	11	9	8	10
1982–83	4,224	1,436	726	6,386	4	9	7	5
1983–84	5,204	1,730	775	7,709	23	20	7	21
1984–85	5,486	1,886	839	8,211	5	9	8	7
1985–86	5,661	2,005	808	8,474	3	6	–4	3
1986–87	5,716	2,079	850	8,645	1	4	5	2
1987–88	5,717	2,187	880	8,784	a	5	4	2
1988–89	7,251	2,721	1,068	11,040	27	24	21	26
1989–90	7,386	3,014	1,168	11,568	2	11	9	5
1990–91	7,997	3,221	1,328	12,547	8	7	14	8
1991–92	8,046	3,321	1,361	12,728	1	3	2	1
1992–93	8,427	2,844	884	12,155	5	–14	–35	–5
1993–94	8,788	2,669	909	12,366	4	–6	3	2
1994–95	9,507	1,932	1,152	13,564	8	10	24	10
1995–96	9,823	2,898	1,156	13,877	3	–1	3	2
1996–97	10,302	2,977	1,030	14,309	5	3	–11	3
1997–98	10,902	3,914	1,373	16,188	6	31	33	13
1998–99	11,354	4,112	1,332	16,798	4	5	–3	4
1999–00	11,593	4,798	1,679	18,070	2	17	26	8
2000–01	10,920	4,215	1,568	16,703	6	–12	–7	–8
2001–02	11,310	4,383	1,667	17,360	6	6	9	6
2002–03	10,464	4,019	1,601	16,094	–7	–8	–4	–7
2003–04	11,077	4,124	1,598	16,799	6	3	–1	4
2004–05	11,714	4,624	2,061	18,398	6	12	29	9

Source: Fertiliser Statistics, 2004–05, p. 181.

increasing the domestic production capacity substantially.[10] Secondly, a cooperative network for fertiliser distribution was created which was eventually replaced by a more efficient marketing network where cooperative and private trade operated side by side.[11] And thirdly, agronomic research related to developing fertiliser responsive varieties, their adaptation and fertiliser use practices, and their dissemination became an integral part of the overall agricultural research and extension.

Between 1952–53 and 2004–05, fertiliser use increased from 66,000 tonnes to about 18.4 million tonnes of plant nutrients (Table 2.2). In this period, the nutrient use increased by about 353 kg of NPK per annum. This amounts to a growth of about 11 per cent per annum. Although this is no small achievement, per hectare use of 90 kg is still one of the lowest in the world. About 50 per cent of this growth in fertiliser use has occurred only around the last 15 years between 1987–88 and 2000–02. However, there are wide variations in year-to-year changes in fertiliser use. In the last 53 years ending 2004–05, there were five occasions when fertiliser consumption was lower than the previous year. The rapidly increasing use after mid-1960s could not maintain the same pace in the early 1970s. The stagnation of fertiliser use during early 1970s was followed by a reduction of about 9 per cent in 1974–75. After this stagnation and decline, fertiliser use has again picked up at a fast rate, except for the years 1992–93, 2000–01 and 2002–03. Although the growth in fertiliser use has been quite stable in 1900s, the erratic behaviour of phosphate and potash fertiliser is concerning. In the last 13 years ending 2004–05, decline in consumption of P and K has been registered five to six times respectively, while consumption of nitrogenous fertilisers has increased stably except for 2002–03. The decontrolled price environment of fertilisers has seriously affected the price of the two nutrients, P and K, and the economics of the use of these nutrients (Table 2.3). As a result, the growth in fertiliser consumption, thus, has been driven by use of nitrogen. This growth on the one hand has taken place at the cost of balanced fertiliser use, on the other hand with significant geographical disequilibrium.

The growth in fertiliser use has been generated by a few geographical locations pre-1979–80. Punjab, Uttar Pradesh, Andhra Pradesh, Tamil Nadu, and Maharashtra not only accounted for about 60 per cent of total fertiliser consumption

Table 2.3
Economics of Application of N and P_2O_5 on Paddy, Wheat, Gram, and Jowar

Particulars		*1971–72*	*1981–82*	*1991–92*	*1997–98*	*2000–01*
Kg of Wheat required to buy 1 kg of	N	2.64	3.93	2.66	1.56	1.64
	P_2O_5	2.45	4.48	3.03	2.93	2.53
Kg of Gram required to buy 1 kg of	N	—	3.52	1.33	0.98	0.91
	P_2O_5	—	4.02	1.51	1.83	1.40
Kg of Jowar required to buy 1 kg of	N	3.65	4.41	3.24	2.21	2.25
	P_2O_5	3.38	5.03	3.69	4.15	3.47
Kg of Paddy required to buy 1 kg of	N	3.79	4.44	2.89	1.92	1.96
	P_2O_5	3.51	5.07	3.29	3.60	3.03

Source: Fertiliser Statistics, 2002, Fertiliser Association of India, New Delhi.

of the country in 1979–80 but also determined a significant proportion of ups and downs of total fertiliser consumption. These five states accounted for over two-third of total increase in fertiliser use in the country between 1961–62 and 1968–69, about half of the increase in fertiliser use between 1968–69 and 1973–74, about 87 per cent of fall in fertiliser use in 1974–75, and about two-thirds of the increase in fertiliser use between 1974–75 and 1979–80. The pace and the pattern of the geographical growth in fertiliser use have been significantly altered in 1980s and1990s. The growth has not only become broad based on account of Karnataka, Haryana, West Bengal and Bihar equally contributing to the growth, it also shows signs of weakening of the growth in three traditionally high growth states, Punjab, Tamil Nadu and Maharashtra (Table 2.4). What is however, concerning is the growth performance of states like Madhya Pradesh, Rajasthan and Gujarat, which is showing signs of stagnation and decline in recent past.

Table 2.4
Growth of Fertiliser Use in Different States

State	*Average annual change ('000 tonnes of NPK) between*			
	1961–62 & 1979–80	*1980–81 & 1992–93*	*1992–93 & 1996–97*	*1996–97 & 2001–02*
Uttar Pradesh	55	171	118	48
Punjab	37	37	2	21
Tamil Nadu	29	26	–2	33
Andhra Pradesh	27	78	51	82
Maharashtra	22	59	13	64
Gujarat	20	30	19	–13
Karnataka	19	36	3	104
Haryana	12	32	30	34
West Bengal	12	37	33	38
Bihar	9	33	38	39
Madhya Pradesh	8	50	22	–2
Rajasthan	8	29	42	–7
Kerala	4	9	–3	–3
Orissa	3	11	10	15
Himachal Pradesh	2	1	a	1
Jammu & Kashmir	1	2	0	7
Assam	a	1	6	14
All India	267 (3.6)[b]	553 (2.8)	539 (3.5)	599 (2.2)

a Less than 500 tonnes.
b Figures in brackets are annual change in NPK per hectare.

Source: Derived from various issues of Fertiliser Statistics, Fertiliser Association of India, New Delhi.

The same phenomenon could be seen at a still disaggregated level. Sah (1982), analysing concentration of fertiliser use in a few districts, reveals that 82 districts that contributed to 88 per cent of total growth of nitrogen use between 1974–75 and 1976–77 have governed the growth pattern of the country. Incidentally, 62 (80%) cent of these 82 districts are from Uttar Pradesh, Punjab, Andhra Pradesh, Tamil Nadu and Maharashtra.[12] These 62 districts were responsible for about 60 per cent of total increase in nitrogen use between 1960–61 and 1968–69, about 42 per cent of total increase between 1968–69 and 1973–74 and nearly all the fall in nitrogen use between 1973–74 and 1974–75. However, neither all the 82 high growth performance districts are from five high growth, high consumption states nor all of them performed equally well throughout the period. For example, nearly one-fourth of these 82 districts had very poor growth between 1960–61 and 1968–69, and about the same number of districts had negative growth between 1968–69 and 1973–74. This concentration of fertiliser use in few districts though weakening of late, the situation over the year has changed marginally. Less than 25 per cent of districts consume over 50 per cent of fertiliser in the country (FAI, *Fertiliser Statistics*, 2001–02). Is this an indication of widening up of the base that generated the growth?

The quantitative increase in fertiliser use in India has been remarkable except for a short stagnation and fall during 1972–75, 1992–93 and 2000–01. Fertiliser use, in long stride, has shown an increasing trend. This growth is about 6,70,000 tonnes of plant nutrients per annum in 1990s as against a moderate growth of 1,13,000 tonnes between 1952–53 and 1970–71 (Table 2.5). However, even if the fertiliser use continues to increase at the same pace as it has between 1992–93 and 2001–02, the actual fertiliser use would fall short by about 21,30,000 tonnes each year than the required growth to meet the Tenth Plan target. The importance of increasing growth of fertiliser use is obvious by looking at its future estimates developed by different agencies (FAI, 2002–03).

Table 2.5
Growth in Fertiliser Use ('000 tonnes) in India*

Fertiliser	*1952–53 to 1970–71*	*1971–72 to 1991–92*	*1992–93 to 2001–02*	*Required growth to meet the 2011–12 targets***
Nitrogenous (N)	75	337	342	NA
Phosphatic (P)	26	145	234	NA
Potash (K)	12	52	95	NA
Total	113	534	670	2,800

* Linear growth by fitting equation $y = a + b.t$ where 'a' and 'b' are parameters, y is fertiliser use and t is time. In the above equation, parameter 'b' gives the linear increase in fertiliser use '000 tonnes per annum.

** Required growth between 2001–02 and 2011–12 to meet the plan targets. The Ninth Plan has targeted the level of NPK use at 45.48 million tonnes in 2011–12.

Source: Various issues of *Fertiliser Statistics*, Fertiliser Association of India, New Delhi.

A cursory glance on the way the fertiliser use has been increasing of late vis-à-vis the required growth, it is quite clear that the projections are astronomically high as compared to the past.[13] Secondly, it also implies that a very rapid growth is needed in order to meet the agricultural production targets. The projection made by Planning Commission with respect to demand for fertilisers to meet the food and non-food agricultural demand at the end of 2011–12 is 45.48 million tonnes of NPK (Table 2.5). This also means that for next ten years fertiliser use must increase at about 28,00,000 tonnes per annum; a tall order to achieve.

Can the growth of fertiliser use be increased and sustained for a long period? This is important question, for states like Madhya Pradesh have shown signs of stagnation and decline in fertiliser use. This question can be answered satisfactorily if the enquiries are directed towards the forces affecting cultivators' demand for fertiliser use in low consuming areas. This is important because fertiliser use is an individual's decision, and at macro level it is aggregation of all the cultivators' consumption that matters. But, before a fresh investigation is undertaken to identify forces affecting fertiliser use and formulate policies for attaining such rates, it would be useful to (a) analyse fertiliser use across districts in Madhya Pradesh, and (b) review the literature on fertiliser use. This would help in understanding how this issue has been examined by researchers in the past.

Fertiliser Use in Madhya Pradesh

This section tries to examine the variations in fertiliser use across districts in Madhya Pradesh. The analysis is based on secondary data pertaining to per hectare NPK use and its correlates. The district level estimates of fertiliser use are available from two different sources, the Fertiliser Association of India and the estimates of Department of Agriculture, Government of Madhya Pradesh. In the present analysis, estimates available from Fertiliser Association (Western Region) for fertiliser use and factors affecting it like cropped area, irrigation, source of irrigation, spread of HYVs, access to cooperative credit and rainfall are used.

With an average of 56 kg of NPK per hectare in 2003–04, fertiliser use widely varies across districts in Madhya Pradesh. At the one end, there is Indore where per hectare NPK use is 123.5 kg, at the other extreme it is Dindhori district with NPK use of only 3.1 kg per hectare. There are four high fertiliser consuming districts (Indore, Gwalior, Khargone and Dhar) in the state where per hectare NPK use is over 100 kg. On the other hand, there are five districts (Dindhori, Shahdol, Umaria, Sidhi and Mandla) with an average per hectare use of less than 20 kg of NPK. The other noticeable feature of fertiliser use in Madhya Pradesh is that per annum increase is significantly and positively associated with base level of fertiliser use. This means that fertiliser use in Madhya Pradesh increased rapidly in districts where the use to begin with was relatively high (Table 2.7). This growth phenomenon in Madhya Pradesh may be guided by the level of agricultural development of the districts; a relatively rapid increase in fertiliser use may be confined to districts with relatively higher level of agricultural potential. It can be argued that factors like irrigation and its control by farmers, spread of HYVs, access to cooperative credit and level of rainfall may be positively associated with level as well as increase in fertiliser use.

Table 2.6 presents the joint distribution of 45 districts of Madhya Pradesh based on level of per hectare NPK use and levels of irrigation. A cursory glance at the table revels that while a relatively high proportion (53%) of low fertiliser consuming district is also having relatively low level of irrigation, a sizable proportion (53%) of districts among relatively high fertiliser use districts are with higher level of cropped area irrigated. This tabular representation is also supported by statistical significance (Chi-Square = 8.4 at significance level of 0.08 and correlation coefficient = +0.4 at significance level of 0.01), implying that per hectare fertiliser use is significantly and positively associated with irrigation. Districts with higher levels of cropped area irrigated are also using higher levels of NPK per hectare. A similar analysis is attempted to identify other correlates of fertiliser use. The analysis is attempted for both the level of NPK use as well as the per annum increase in it (Table 2.7).

Table 2.6

Association between Fertiliser Use and Level of Irrigation across Districts

Level of fertiliser	*Percentage of cropped area irrigated*			*All*
	Low	*Medium*	*High*	
Up to 45 kg/ha	53	40	7	100 (15)*
45.1 to 67 kg/ha	27	33	40	100 (15)
Over 67 kg/ha	20	27	53	100 (15)

*Number of districts,

Note: chi-square = 8.4 significance 0.08, r = +.4 significance 0.01

It is observed that while level as well as per annum increase in fertiliser use is positively and significantly associated with cooperative credit, base level of fertiliser use and level of irrigation and farmers' control on irrigation influence the level of fertiliser use but not growth in fertiliser use. Similarly, spread of HYVs influences the growth in fertiliser use but does not influence level of fertiliser use. The foregoing analysis revels that while fertiliser use is governed by agricultural development indicators like access to cooperative credit, irrigation, control on irrigation, per annum increase in fertiliser use is governed by access to institutional credit, controlled irrigation and spread of HYVs.

Table 2.7

Correlation of NPK Use Per Hectare with Some Important Factors

Factors	*NPK use per ha*		*Change in NPK use per annum*	
	Chi-square	*Correlation*	*Chi-square*	*Correlation*
Cooperative Credit	23.6 (0.02)	0.63 (0.00)	12.8 (0.01)	0.47 (0.00)
Net Area Sown	3.8 (NS)	0.23 (NS)	5.6 (NS)	0.07 (NS)
Percentage Area Irrigated	8.4 (0.08)	0.4 (0.01)	4.4 (NS)	0.23 (NS)
Percentage HYV Area	3.4 (NS)	0.23 (NS)	5.9 (0.10)	0.23 (0.10)
Percentage Canal Irrigation	6.8 (0.10)	–0.24 (0.10)	13.5 (0.01)	–0.37(0.01)
Rainfall	4.3 (NS)	0.06 (NS)	4.2 (NS)	0.1 (NS)
Base Fertiliser Use	29.6 (0.00)	0.67 (0.00)	15.2 (0.00)	0.3 (0.02)

These explanatory factors taken together are able to explain about 65 per cent variations in per hectare NPK use across districts in Madhya Pradesh (Table 2.8). However, the identified factors are able to explain about 38 per cent variations in per annum increase in fertiliser use. Interestingly, increase in proportion of canal irrigation in total irrigation is not only reducing the level of fertiliser use but also decreasing the growth in fertiliser use as well. This implies that fertiliser use as well as per annum increase in it is faster in districts that do not depend on canals for irrigation. In fact, dynamism in Madhya Pradesh agriculture was governed by well irrigation that is dependable and controlled by farmers. On the other hand, canal irrigation is controlled by state and is highly undependable as a source of irrigation. Consequently, deep bore wells have mushroomed across states pushing water tables alarmingly low.

Table 2.8

Linear Regression Coefficients of Factor Explaining Variation in Fertiliser Use

Factors	*NPK use per hectare*	*Per annum change NPK use*
Per cent Irrigation	0.437 (0.02)[a]	—
Cooperative Credit	0.266 (0.00)	9.126 (0.01)
Per cent Canal Irrigation	–0.494 (0.00)	–9.170 (0.01)
Percentage of HYV Area	—	6.416 (0.02)
Constant[b]	26.9	8.8
R bar square	0.65	0.38
F	28 (0.00)	9.9 (0.00)

a Figure in parentheses are level of significance
b Capture the base level of NPK use per hectare also.

To sum up, the cross-sectional analysis of fertiliser use in districts of Madhya Pradesh reveals that forces like irrigation and control on it, spread of HYVs and availability of cooperative credit are important factors that govern fertiliser use in Madhya Pradesh. These facets of dynamism in fertiliser use are not new to knowledgeable scholars of fertiliser industry. But these facets are a fair indicator towards the processes that need to be included in a study that explores the level and increase in fertiliser use in a low potential area. The analysis in subsequent chapters shall unfold these processes in tribal south-western Madhya Pradesh.

Review of Literature

Adoption and Diffusion

A number of research efforts were directed in early 1960s towards identifying factors affecting fertiliser use at farm level. Some of these identified farmers' characteristics associated with fertiliser use[14] and some were addressed to identify farm enterprise characteristics associated with adoption and diffusion of fertilisers.[15] The first group of studies emphasised social forces like caste, education, age, status of farmer, contact with extension agency, etc. as factors responsible for fertiliser use. These studies could be criticised for two failings. Firstly, these researchers were trying to find answers for an economic phenomenon in non-economic factors. Secondly, and more importantly, these researchers were explaining a dynamic phenomenon by static factors. The second group of studies looked upon farm enterprise as the unit of observation and explained fertiliser use on the basis of economic variables. Since these researchers looked at farm enterprise, fertiliser use on different crops was their main concern.[16] These studies found differences in cropping pattern and irrigated area as the main forces behind the differences in fertiliser use across the farms and the crops.

Desai, Chary and Bandopadhyay looked into the dynamics of fertiliser use in a continuous cross-section in Andhra Pradesh between 1967–68 and 1969–70.[17] Their findings throw valuable insight on the diffusion as well as adoption processes. They observed that it took over 15 years for some cultivators to adopt fertilisers after they first heard about it. There was no significant association between gestation period for adoption with variables like formal education, caste, tenancy status, and size of farm. The reason for late adoption of fertiliser use was dominance of low value crops in the cropping pattern of the non-adopters. While almost all users diffused fertilisers over time, such diffusion was slow on crops such as jowar, bajra, varagu and pulses even after a number of years.

The above studies highlighted the importance of economic variables in explaining fertiliser use. Some of these are pioneering in the sense that they indicated that the important question is why cultivators use fertilisers and not who uses it. But there are limitations too in these efforts: (1) These studies are based on a cross-section at a point of time and hence lack dynamism. (2) They were conducted in early 1960s and since then fertiliser use has grown manifold. In fact, the bulk of the growth in fertiliser use, as the overview in the preceding section has shown, has occurred only after mid-1960s, more specifically after 1986–87. (3) Diffusion of fertiliser use was not complete at the time these studies were conducted. (4) After these studies, specifically after 1992, the technological and price environment has undergone substantial change. Consequently, forces affecting fertiliser use may have changed.

Response Functions

The economic rationale to some of the above studies was provided by the fertiliser response experiments conducted on cultivator's fields on different crops during pre-HYV period.[18] Introduction of HYVs brought optimism about the response superiority of HYVs over traditional varieties and economic benefits of balanced fertilisation.[19] The main issue was to establish or demolish the wisdom:[20] (1) that HYVs give higher response than the traditional varieties, (2) that full benefit of HYVs can be achieved when synergistic potential of package of inputs is exploited, and (3) that among the inputs most important are fertilisers and irrigation. This resulted in a policy towards concentration of fertiliser use on HYVs with assured water supply and intensive use of complementary inputs.[21] This issue consequently became a source of controversy for what ought to be the right strategy.[22] The above controversy was seen by Herat and Mellor even in early 1960s when they compared response functions of rice in Orissa (India) and Arkenaes (USA).[23] They found that US rice had about three times more agronomic capacity and about four times more agro-economic capacity for using fertilisers than the Orissa rice. However, such findings did not find place in the economist's kit to formulate policies towards fertiliser use till very late.[24]

The use of fertiliser experiment results in different agro-climatic zones, became the base for recommendations about fertiliser use. Such recommendations, however, were evolved in relatively stable and favourable fertiliser prices of 1950s and 1960s. With the oil crisis of early 1970s, fertiliser price increased by about 100 per cent in 1974–75 over the previous year. This increase was considered by the fertiliser industry as the foremost factor influencing fertiliser use.

Fertiliser industry was trying to establish that high prices of fertilisers have resulted in stagnation and decline in fertiliser use during 1974–76.[25] Support to their views was again sought from response-functions of different crops under changing price environment. This was indicative of normative view of economic optimum rate of fertilisation at two different fertiliser-crop price ratios. Use of this type of ratio in farm level observation has limitations. This would indicate that at deteriorated fertiliser-crop price ratio, the optimum rate of fertiliser use would decline. In actual practice, there could be an increase in fertiliser use even in deteriorating price environment. This is because at the farm level, crop and fertiliser price would individually decide the economic returns rather than jointly through price ratio.

On the other hand, the evidences on change in actual use of fertilisers on farms in deteriorated price environment are mixed.[26] There are evidences to show that fertiliser use has increased but there are also crops and locations where fertiliser use has decreased. These evidences are both for foodgrains and non-foodgrains and for HYVs as well as traditional varieties.

Fertilise Use Practices and Agronomic Variables

The loss of applied fertiliser nutrients of soil has always been an agronomist's concern[27] and the oil crisis of 1970s induced the economist's attention towards this issue. There are two main aspects of this issue. First is soil-crop-fertiliser management, and second is time and method of fertiliser use. The yield response to fertiliser use can be improved over the general recommendations if the soil fertility status of the field is known beforehand.[28] It is now established that the economic rate of fertiliser use is negatively associated with the fertility status of the soil.[29] Although, the economic gains of soil test are well established, Babaria's findings show that the follow-up of results at the farm level was discouraging. However, nearly half of the cultivators in his sample believed that fertiliser use should be based on soil tests.[30]

Yield response to fertiliser application also improves by making the nutrients available to the plants at the right time. Experiments have shown that split application of nitrogen on paddy and wheat is more efficient than a single basal dressing.[31] More recently, the experiments on new fertiliser material have gained popularity. Research experiments with nitrification inhibitor and slow releasing nitrogenous fertilisers revealed that they are more efficient in making the nutrients available to plants, thereby improving the yield gains compared to ordinary nitrogenous fertilisers.[32] Such experiments were also successfully conducted using indigenous and relatively cheap materials like coated urea, *neem* extracted urea, and tar treated urea.[33]

The effect of these findings is yet to be felt in the actual fertiliser use pattern. This is so because first of all the existing data do not permit to investigate the effect of these factors, and secondly, these agronomic practices themselves have to find wide economic recognition on diverse conditions over different agro-climatic zones.

Dynamics of Fertiliser Use at Macro Level

The pattern of fertiliser use at the district level reveals concentration in selected locations. Desai's findings reveal that less than one-third of total districts in India accounted for over 70 per cent of total fertiliser use during any year between 1960–61 and 1968–69.[34] A large number of high fertiliser consuming districts were from Punjab, Haryana, Uttar Pradesh, Maharashtra, Andhra Pradesh and Tamil Nadu. Desai tried to explain this pattern of growth in fertiliser use by using micro-level understanding. He concluded that the variations in the growth rates were mainly due to differences among districts with respect to irrigation, cropping pattern, and diffusion of HYVs.[35]

Analysing influence of demand versus supply factors in fertiliser use at macro level, Nampoothiri and Desai (1994) have suggested that the relative importance of different facets changes at different steps of growth in fertiliser use. These researchers conclude that it is important to understand the process on both demand-side and supply-side that has governed the pace and pattern of growth in fertiliser use, and also the role played by the state government. Unirrigated areas can respond to supply processes and the interactions between outlets, credit and kind component in crop loans, supply variables like distribution networks, share of cooperatives and the ground level demand variables like irrigation, rainfall, cropping pattern, spread of HYVs and credit. The overview of growth in fertiliser use presented in the beginning of this chapter reveals that the concentration of fertiliser use still persists among states as well as districts. The important question is, are the same forces still operating in influencing fertiliser use, which were working during the mid-1980s and 1990s? Other researchers have also examined the growth of fertiliser use at macro level. Parikh, who uses continuous cross-section of states, has explained that in the long run the reduction of real price of fertiliser and increasing the proportion of irrigated area significantly increased the consumption of fertilisers.[36] Employing covariance analyses, he, however, finds that in the short-run economic variables have little effect on fertiliser use. Maharaja has tried to explain growth of fertiliser use in Gujarat during 1960s.[37] He finds that spread of irrigation and HYVs were two major factors determining the growth in fertiliser use.

The methodology employed by these researchers has serious limitations. The demand models used by them are derived demand functions in which production function variables are replaced by instrumental variables like irrigation, HYVs, etc. This means that their methodology ignores the differences in production response over location and time. Secondly, this approach totally overlooks the relevance of diffusion of fertilisers; in a given agronomic environment there exists a potential for fertiliser use. Irrespective of changes in instrumental variables the actual use grows towards this potential by a simple diffusion process.

There is another dimension to policy that governs economic environment in which fertiliser use has grown, the pricing policy of fertilisers. Government policies on fertiliser sector have been uncertain and inconsistent in last five decades. This, of late, has resulted in growing fiscal burden as well as distortions in the use pattern. Although, the subsidy for fertiliser use in India started with the basic objective of rapid agriculture development by promoting adoption of new seed-fertiliser technology, especially for risk-averse farmers, over the years the mounting burden of subsidy led to steps in mid-1990s to rationalise the subsidy. Serious researcher, though questioning the relevance of subsidies even earlier (Sagar, 1991), the issue started gaining political dimensions only after 1992. The subsidy at one level had encouraged malpractices in order to corner a share of fertiliser subsidy by industry, called gold plating (see, for details, Venkateshwarlu and Sen, 2000), at the other level, the price distortions have encouraged inefficient fertiliser use, whereby NPK ratios distorted owing to relatively high prices of phosphatic and potash fertilisers vis-à-vis nitrogen, and severe regional disequilibria in fertiliser use and iniquitous benefits to rich class of farmers (Gulati and Sharma, 1995; Vaidyanathan, 2002; Reddy and Deshpande, 1992; Rao and Srivastava, 2002). Singh (2004) has argued that implications of this rationalisation are still unclear, as a uniform approach to reducing fertiliser subsidy is not justified. Instead a well-thought out, properly sequenced, gradually and regionally differentiated approach to subsidy reduction needs to be adopted.

Fertiliser subsidy post-1992 has become a contentious issue especially for locations where fertiliser use is near optimum levels. While, urea is subsidised at present under fertiliser retention price systems (RPS), phosphate and potash have been decontrolled. Industry, fearing a slide in consumption in case subsidy is withdrawn, insists a subsidised pricing environment. But the government is clear in slowly bringing the subsidy to a manageable level for, subsidy in last two decades ending 2000 has reached a staggering level of Rs 133 billion (Gulati and Naryanan, 2003). Fertiliser subsidy, through RPS, has become a burden on exchequer. But more disturbingly, RPS provides, as mentioned above, an avenue to fertiliser industry to corner a substantial proportion of the subsidy. Gulati and Naryanan (2003) estimated that about a third of the subsidy is cornered by the industry and only two-thirds of the subsidy is passed to farmers. Despite this gold plating by the industry, Indian farmers have paid lower domestic prices than what they would have paid if there were free trade and no subsidy to farmers.

Dynamics of Fertiliser Use at Micro Level

Using the concept of potential, Desai (1969; 1973) argues that despite an optimum level of use at which net returns to farmers are optimised, fertiliser use at the micro level widely varies both in terms of area fertilised as well as rates of nutrient application. Empirical research relates this variation to both yield response and the local agro-climatic conditions in which the fertiliser is applied. These findings reveal importance of following factors in governing fertiliser use at micro level: rainfall (Cummings and Ray, 1969), irrigation and its quality (Desai, 1973; Bliss and Stern, 1982; NCAER, 1978; Parikh and Mosely, 1986; Nagraj, 1983), cropping pattern and varieties (Rao, 1976; Parthasarthi, 1978) relative price of fertiliser (Desai, 1986; Parikh, 1965); availability of credit (Desai and Stone, 1987), tenancy and land holdings (Sivakumar, 1986) and management practices of farmers (Mohanam, 1990).

The response of yield to fertiliser use, as observed on fertiliser experiment stations, sets the limits for the potential for fertiliser use. Findings for dry crops: oilseeds, cotton, maize, and pulse in dry tribal belt show that the economics of available seed-fertiliser technology is conducive to wider adoption only in the case of phosphate rather than nitrogen.[38] Some farmers are using only nitrogenous fertilisers, urea mainly (Sah and Shah, 1993). Sometimes the basal doses are applied even after sowing the leguminous crops only when it becomes certain that the rainfall is normal. Findings about the response environment for irrigated crops, based on ABC trials, suggest[39] that available seed-fertiliser technology is (i) highly conducive to overuse of fertilisers in the case of oilseeds and wheat (ii) such that the level of fertiliser use shall be close to the recommended rates for gram and cotton; and (iii) such that fertiliser use shall be at a low level for crops like pulse, *jowar* and maize. Since fertiliser use at micro level is consistent with what might be expected on the basis of results obtained at research experiments (Sah and Shah, 1990; 1994), the key to future growth in fertiliser use lies in (i) improving yield response of (a) rainfed cotton to added nitrogen, (b) rainfed oilseeds and pulses to added phosphate, and (ii) in improving the efficiency of fertiliser use in irrigated wheat.

A number of studies, as stated earlier, have identified factors affecting demand for fertilisers at farm level. Some[40] identified farmers' characteristics associated with fertiliser use like social categories including caste, education, status and contact with extension agencies. Some others[41] address to identifying farm enterprise characteristics associated with adoption and diffusion of fertilisers. It were, however, the efforts of Minhas and Srinivasan (1966), Cummings Jr., et al. (1968), Parikh (1978) and Vaidyanathan (1978)[42] that established the conventional wisdom that fertiliser-irrigation synergy is the main form of dynamism of fertiliser use both before and after the introduction of HYVs. The revealing insight about adoption of fertiliser and its diffusion, nonetheless, were provided by Desai (1969) and Sah (1982) who concluded that (a) the size and certainty of return on fertiliser use govern the pattern of use at the micro level; and (b) at the macro level, the actual fertiliser use, however, moves asymptotically towards a potential, and that the growth in fertiliser use is rapid until crops with relatively high and certain returns are fertilised near optimal rate, thereafter the growth slows down because the untapped potential consists of crops on which the returns are low and uncertain. It can, however, be argued that there could be a mismatch between farmers' perceptions about the optimal doses of fertilisers and the recommended rates.

Variations in fertiliser use with reference to the recommended rates are mainly guided by farmers' perceptions that if they accept the recommendation, the yield will be lower than their present yield. But there is little clarity at present as to how these perceptions are formed and more importantly how they can be influenced. The widespread misinformation on the effect of phosphate in irrigated areas (Sah and Shah, 1992) and the lack of effective soil-moisture conservation techniques in dry farming zone may also be the reasons for wrong perceptions (Shah and Sah, 1995). Moreover, constrained availability of working capital in this tribal zone and the substitution of fertilisers by labour may be another cause of low use of fertilisers on different crops. There is also a lack of appreciation of the positive role the cooperative marketing network has played in creating demand for fertilisers.

Summing up

The above review provides enough insight on forces governing fertiliser use at micro level, but fails to explain the growth of fertiliser use at macro level. This is primarily because micro level findings identify irrigation and cropping pattern as the main variables influencing fertiliser use among different crops and farms. The same variables are used in macro models. In the process, the importance of production response differences and potential for fertiliser use was ignored. Similarly, the investigation of response functions at micro level was dominated by issues like what ought to be done rather than trying to understand why cultivators used fertilisers the way they used.

Apart from these limitations, the studies also suffer due to the fact that they were carried out when fertiliser use was low, price environment was favourable to farmers and a sizeable proportion of potential for fertiliser use was untapped. Since then not only fertiliser use has grown substantially but the environment in which it was growing has also undergone considerable changes. Thus, the need to re-examine the factors governing fertiliser use—especially its adoption, diffusion and continuous use—in areas like Western tribal belt of Madhya Pradesh, which are still outside the preview of growth becomes vital after this review.

Notes

1. J.A. Voelcker, Report on the Improvement of Indian Agriculture, Eyre and Spottiswood, London, 1893, p. 41; quoted by Gunvant M. Desai, "Growth of Fertiliser Use in Indian Agriculture: Past Trends and Future Demand", Cornell International Development Bulletin 18, USA, 1971, pp. 8–9.
2. The Royal Commission on Agriculture 1928 reveals, "It is hardly necessary to point out that the use of nitrogenous or other artificial fertilisers is not profitable in all conditions. Where crop production is limited by small rainfall, the annual addition of chemical nitrogen to soil as the result of natural process may be sufficient to meet the need of the crop". Reproduced in *Report of the National Commission on Agriculture 1976*, GDI, Ministry of Agriculture and Irrigation, Part X (Inputs), 1976.
3. Planning Commission, *First Five-Year Plan*, Government of India, New Delhi, 1952, pp. 254–59.
4. Planning Commission, *Second Five-Year Plan*, Government of India, New Delhi, 1956, p. 260.
5. Planning Commission, *Third Five-Year Plan*, Government of India, New Delhi, 1961, pp. 304–13.
6. See, *Approach to Agricultural Development in Fourth Five-Year Plan*, Government of India, Ministry of Food and Agriculture, New Delhi, 1964, pp. 24–28; Planning Commission, *Fourth Five-Year Plan*: A Draft Outline, Government of India, New Delhi, 1966, pp. 173, 196–97, 251–53; Planning Commission, *Draft Fifth Five-Year Plan*, 1974–*79*, Vol. 11, Government of India, New Delhi, 1973, p. 12; India, Planning Commission, *Draft Sixth Five-Year Plan*, 1978–63, Government of India, New Delhi, 1978, pp. 141–42; Planning Commission, *Sixth Five-Year Plan*, 1980–*85*, Government of India, New Delhi, 1981, p. 106.
7. Planning Commission, *Fourth Five-Year Plan*, 1969–74 (Revised), Government of India, New Delhi, 1970, p.113.

8. *Draft Sixth Five-Year Plan*, 1978–83, op. cit., p. 142.
9. The plan targets of fertiliser consumption were 8 million tonnes. This was subsequently revised and reduced to 4.8 million tonnes. The revised target was even lower than the Fourth Plan target (Sah, 1982).
10. The dependence on imports has reduced significantly over the years. For example, from 80 per cent of total supply on the eve of independence, the import content has reduced to about 30–35 per cent in the late 1970s. But in absolute terms the imports are about 1.5 to 1.6 million tonnes which is about 3.5 to 4 times higher than what they were in early 1960s. Even in 2001–02, about 14 per cent of total fertiliser consumed in the country was imported (FAI, Fertiliser Statistics, various issues).
11. This was introduced for creating adequate atmosphere to attract private and foreign capital in fertiliser industry. See Ministry of Food and Agriculture, Report of the Committee on Fertilisers, GoI, New Delhi, 1965, p. 51.
12. These districts are Cudappah, Karimnagar, Kurnool, Mehaboob Nagar, Medak, Neloore, Viahakapatnam, East Godavari, Guntur, Krishna, Nizamabad, and Warangal in Andhra Pradesh; Nagpur, Satara, Wardha, Dhulia, Jalgaon, Kolhapur, Nasik, Sangli in Maharashtra; Jalandhar, Kapurthala, Amritsar, Bhatinda, Ferozepur, Gurdaspur, Ludhiana and Patiala in Punjab; Chingleput and Coimbatore in Tamil Nadu; and Agra, Mainpuri, Etah, Bareilly, Bijnor, Pilibhit, Rampur, Farukhabad, Etawah, Allahabad, Lucknow, Kheri, Gonda, Behraich, Saharanpur, Muzaffarnagar, Meerut, Bulandshahr, Aligarh, Moradabad, Shahjahanpur, Nainital, Barabanki, Gorakhpur, Deoria, Basti, Azamgarh, Varanasi, Mirzapur, Jaunpur, Gazipur, Ballia, Faizabad, and Badaun in Uttar Pradesh.
13. Gunvant M. Desai, *Growth of Fertiliser Use*, op. cit., p. 16; and Maharaja, *Demand for Fertilisers*, Good Company, 1975, p. 64.
14. M.S. Shetty, "Inter Farm Rates of Technological Diffusion in Indian Agriculture", *Indian Journal of Agricultural Economics*, Vol. 21 (1), January-March 1966, pp. 189–98; V.N. Mishra, "The Effect of Institutional Factors on the Use of Fertilisers", *Indian Cooperative Review*, Vol. 4 (4), 1967, pp. 120–22; M.S. Shetty, "Factor Analysis of Use of Fertilisers by Farmers", *Indian Journal at Agricultural Economics*, Vol. 24 (1), January-March 1969, pp. 50–61; D.K.

Desai and B.M. Sharma, "Technological Changes and Rates of Diffusion," *Indian Journal of Agricultural Economics*, Vol. 21 (1), January-March 1966, pp. 141–46; K.M. Chaudhari and M. Maharaja, "Acceptance of Improved Practices and their Diffusion among Wheat Growers in Pali, Rajasthan", *Indian Journal of Agricultural Economics*, Vol. 21 (1), pp. 161–65.

15. K.S. Rao, "Some Economic Aspects of Fertiliser Use in India", *Agricultural Situation in India*, Vol. 15 (12), March 1962, pp. 1215–16; R. Rajagopalan, "Change in the Application of Fertilisers and Economics of Fertiliser Use", *Agricultural Situation in India*, Vol. 18 (8), November 1963, pp. 515–21; V.G. Panse and D. Singh, "Promotion and Assessment of Technological Changes in Indian Agriculture", *Indian Journal of Agricultural Economics*, Vol. 21 (1), January-March 1966, pp. 121–31; D.K. Desai and B.M. Sharma, op. cit., pp. 147–54; Gunvant M. Desai, "Growth of Fertiliser Use in Indian Agriculture: Post Trends and Future Demand", Cornell International Agricultural Development Bulletin No. 18, 1971, pp. 44–56; Gunvant M. Desai, P.N. Chary and S.C. Bandopadhyay, *Dynamics of Growth in Fertiliser Use at Micro Level*, Indian Institute of Management, Ahmedabad, 1979.
16. Gunvant M. Desai, *Growth of Fertiliser Use,* op. cit.
17. Gunvant M. Desai, P.N. Chary and S.C. Bandopadhyay, op. cit., pp. 91–102.
18. A summary of these experiment results was used by Hopper to develop yardsticks for fertilisers, See W.D. Hopper, "Planning Yardsticks for Fertiliser and Irrigation", *Agricultural Situation in India*, Vol. 20 (6), September 1965, pp. 463–77. The response functions developed by Hopper were used by Gunvant M. Desai to explain why fertiliser use varies among different crops.
19. J.S. Kanwar et al., "Balanced Fertiliser Use for Maximising Returns from Wheat on Cultivators' Fields", *Fertiliser News*, Vol. 17 (11), November 1972, pp. 19–30; J.S. Kanwar et al., "Economics of Fertiliser Use in Rice in Farmers' Fields", *Fertiliser News*, Vol.18 (1), January 1973, pp. 71–88; J.S. Kanwar, "Are Fertilisers Application to Jowar, Maize and Bajra Economical?", *Fertilizer News*, Vol. 18 (7), July 1973, pp. 19–28.
20. B.S. Minhas and T.N. Srinivasan, "New Agricultural Strategy Analysed", *Yojana*, January 16, 1966, pp. 13–18; R.W. Cumming et al., "New Agricultural Strategy Revisited", *Economic and*

Political Weekly, Vol. 3 (43), October 26, 1968; Krit Parikh, "HYVs and Fertilisers: Synergy and Substitution", *Economic and Political Weekly*, Vol. 12 (12), March 25, 1978, pp. A2–8; A. Vaidyanathan, "HYVs and Fertilisers: Synergy and Substitution (A Comment)", *Economic and Political Weekly*, Vol. 13 (25), January 1978, pp. 1031–35.

21. Planning Commission, *Fourth Five-Year Plan, 1969–74*, Government of India, New Delhi, pp. 130–31; Reorientation Programme of Agricultural Production, op. cit., pp. 5–12.
22. Parikh's research in this context has far-reaching consequences. Using Fertiliser Trial data on 15,000 fields he has challenged the conventional wisdom and recommended the fertiliser need not be concentrated either on irrigated land or on HYVs. Fertiliser use should be based on local conditions and response of available varieties rather than general principle of synergy. See Parikh, op. cit..
23. R.W. Herdt and J.W. Mellor, "The Contrasting Response of Rice to Nitrogen: India and United States", *Journal of Farm Economics*, Vol. 46 (1), February 1964, pp. 150–61.
24. Mohinder S. Mudahar, "Needed Information and Economic Analysis for Fertiliser Policy Formulation", *Indian Journal of Agricultural Economics*, Vol. 33 (3), July-September 1978, pp. 40–67.
25. Issues of *Fertiliser News*, published monthly by the Fertiliser Association of India, after oil crisis of mid-1970s extensively carried editorials and tables establishing that economics of fertiliser use has become less attractive. Issues like price support and subsidy developed after oil crisis.
26. See papers presented on the topic "Input Price Profitability and Production" and Gunvant M. Desai's Rapporteur's Report in *Indian Journal of Agricultural Economics*, Vol. 31 (3), July-September 1976, pp. 40–47
27. Such losses deprive the soil of plant nutrients through leaching, violatilisation, run off and locking of nutrients in soil. See H.L.S. Tandon, "Dynamics of Fertiliser Nitrogen in India: Usages, Transformation and Crop Removal", *Fertiliser News*, Vol. 19 (7), July 1974, pp. 3–12; H.L.S. Tandon, "Dynamics of Fertiliser Nitrogen in India: Efficiency Losses and Long Term Effects", *Fertiliser News*, Vol. 9 (8), September 1974, pp. 14–21.

28. B. Ramamoorthy and U.V. Pathak, "Soil Fertility Evaluation Key to Targeted Yield", *Indian Farming*, Vol. 18 (12), March 1968, p.29; B. Ramamoorthy et al., "Soil Testing: A Base for Efficient Fertiliser Use for Hybrid Bajra in Draught Allocated Area", *Fertiliser News*, Vol. 14 (8), August 1969, p. 26; Also see S.P. Dhnu et al., "Experience in Farmers' Fields: Soil Testing and Fertiliser Use" (in eds) FAI-FAO Seminar on Optimising Agricultural Production Under Limited Availability of Fertiliser, FAI, 1975, pp. 353–70.
29. J.S. Kanwar, "Coordinated Research in Soil Productivity and Water Management", *Fertiliser News*, Vol. 17 (6), June 1972, pp. 13–16.
30. Babaria finds that only 4 per cent of the cultivators, who have examined their soils, based their fertiliser use on these results. And about 93 per cent of cultivators did not even receive the results of their soil test. These findings are based on a survey conducted in 1973–74 in Surat (Gujarat). See C.H. Babaria, Economics of Soil Testing Services in Gujarat, Ph.D. Thesis submitted to Sardar Patel University, Vallabh Vidya Nagar, 1977, pp. 157–60 and 177–78.
31. Hten Have, "Optimising Time of Nitrogen Application for Transplanted Rice", *Fertiliser News*, Vol. 16 (2), February 1971, p. 12; Ambika Singh and N. Singh (eds.), "Time and Method of Fertiliser Application under Limited Availability" in FAI-FAO Seminar, op. cit., pp. 299–300.
32. Rajendra Prashad (ed.), "Research on Nitrification Inhibitor and Slow Release Nitrogen in India: Use and Problems" (in eds), FAI-FAD Seminar, op. cit., pp. 167–85.
33. S.N. Bains et al., "Use of Indigenous Material to Enhance the Efficiency of Fertiliser Nitrogen for Rice", *Fertiliser News*. Vol. 19 (3), March 1974, pp. 30–52.
34. Gunvant M. Desai and Gurdev Singh, *Growth of Fertiliser Use in District of India: Performance and Policy Implications*, Indian Institute of Management, Ahmedabad, 1973, pp. 26–29.
35. Ibid., Chapter 4.
36. Ashok Parikh, "Consumption of Nitrogen Fertilisers: A Continuous Cross-Section Study and Covariance Analysis", *Indian Economic Journal*, Vol. 14 (2), October-December 1966, pp. 258–74.

37. M.H. Maharaja, *Demand for Fertilisers: An Analysis of Factors Affecting Demand and Estimation of Future Demand,* Good Company, Baroda, 1975, pp. 113–36.
38. Based on discussions with farmers and input dealers. Also see Shah A. and D.C. Sah (1993), "Dry Land Farming Under Changing Resource Environment", *Arth Vijnan,* Vol. 35 (9).
39. The extension wing of GSFC and IFFCO believe that the misperception about phosphate and potash use is widely prevalent in the farming community all over the country.
    Also See D.C. Sah and A. Shah, Soil Testing Facilities and Activities in Gujarat, Gujarat Institute of Area Planning, Ahmedabad (Mimeo), 1990; D.C. Sah and A. Shah, "Incentives for adopting Soil Test based Fertiliser Use", *Indian Journal of Agricultural Economics,* 47 (4), October-December, 1992; D.C. Sah and A. Shah, "Farmers Fertiliser Practices and Soil Testing Services", in G.M. Desai and A. Vaidyanathan (eds.), *Strategic Issues in Future Growth of Fertiliser Use in India,* ICAR India and IFPRI Washington, 1994.
40. See, Footnote 14.
41. See, Footnote 15.
42. B.S. Minhas and T.N. Srinivasan, "New Agricultural Strategy Analysed", *Yojna,* January 16, 1966; Cummings R.W., "New Agricultural Strategy Revisited", *Economic and Political Weekly,* 3(43), 1968; Parikh K., "HYVs and Fertilisers: Synergy and Substitution", *Economic and Political Weekly,* 12 (12), 1978; Vaidyanathan A., "HYVs and Fertilisers, Synergy and Substitution: A Comment", *Economic and Political Weekly,* 13(25), 1978.

Chapter 3

Understanding Dynamics in Fertiliser Use: Conceptual Issues

The objective of this chapter is two-fold: (a) to develop an analytical frame for identifying sources of growth in fertiliser use at the micro level between three points of time; and (b) to present data and methods used in the analysis. The growth of fertiliser use could be seen with respect to adoption, discontinuation or continuous user of fertiliser on a crop, for the usual two-fold classification of cultivators into users and non-users could be misleading because each of this category consists of sub-categories. Equally important is to understand the source of growth in fertiliser use of continuous users. The relative importance of various sources of growth in fertiliser use in a crop would provide understanding on forces that are driving or constraining the growth of fertiliser use in areas where fertiliser use is low.

Conceptual Framework

The scope for fertiliser diffusion in an economy is determined by development and interaction among four sets of processes (Desai, 1982) which influence: (i) agronomic potential for fertiliser use; (ii) conversion of potential into farmers effective demand; (iii) growth of aggregate supply of fertilisers; and (iv) availability of fertilisers at the right time and place. The first process refers to the physical response function and the price environment that together determine the agro-economic potential for fertiliser use. The physical response function depends on factors such as soil, climatic environment, cropping pattern, genetic characteristics of crops, the use of other inputs, etc. Given the physical response function, relative prices of crops and fertilisers would determine the economic level of fertiliser use. If the relative prices remain more or less unchanged, the first process would focus on increasing the agronomic potential for fertiliser use. Improving irrigation or soil conditions on the one hand and strengthening the R&D activities with respect to various farm inputs as well as farm practices on the other could attain such an increase.

The second process relates to the farmers' effective demand that could be influenced through an array of support mechanisms, which help them perceive the economic benefits due to fertiliser use. The main thrust of the strategy would be to raise fertiliser use not only by influencing the motivational characteristics of the potential adopters, but also by providing relevant information as well as other support services. This would include location-specific research and extension, supply of complementary inputs such as seeds, pesticides, manure, development of output markets, and sales promotion campaigns, etc. Micro-level agencies such as farmers' cooperatives, village level extension workers, and sales outlets of manufacturers can play an important role in developing this process. The third process deals with the aggregate supply of fertilisers. It aims at sustaining the growth of fertiliser supply through domestic production and imports. Relevant strategy would be to create a policy environment in which domestic production could be sustained through efficient growth and utilisation of production capacity. Similarly, cost-effective import management would be another important feature of the third process. By and large, the development of this process will be determined by the nature of state intervention. The fourth process focuses on the development of the distribution system to ensure availability of fertilisers at the right time and right place. Establishment of sales outlets, storage facilities, training of the staff employed at the retail outlets in the procurement of fertilisers, and efficient transportation of fertilisers are the main components of this process. This would again involve the development of local level agencies to meet the location-specific requirements.

The first and third processes relate to the macro-level environment and are influenced directly by public policies. The micro-level agencies may contribute mainly in the second and the fourth processes, i.e., converting the potential into effective demand and ensuring smooth flow of fertilisers. Therefore, the evaluation of micro-level diffusion and marketing strategies should focus on the latter two processes.

In a predominantly market-oriented economy, the second and the fourth processes are generally developed under the marketing strategies adopted by the manufacturers. Since the nature and extent of industry's interventions are primarily guided by the profit motive, it is likely that the industry-sponsored agencies may concentrate on a few locations, and on activities that serve their commercial interest. On the other hand, in a quasi-planned economy, the primary responsibility for developing these processes would rest with the government or government-sponsored agencies, which might aim at fulfilling certain national goals, e.g., equity or regional balance etc., along with or through the diffusion of fertilisers per se. Thus, in an economy where fertiliser markets are predominantly administered, these two processes are likely to be significantly influenced by the development objectives and government policies.

In India, these processes have so far been regulated through various policy measures such as institutional support for R&D and extension services, soil testing facilities, development of cooperative credit as well as cooperative marketing network, supply of complementary inputs, market allocation and fixing of distributors' margins. All these factors may not only influence the motivation of decision-makers in the industry but also the nature of their activities. It can be argued that given the extensive policy interventions, the industry and other micro-level agencies could play only a limited role in these two processes. However, any efforts of the industry in the area of creating a linkage between the overall support system created by the government and the farmers would facilitate and enhance the long-term growth of the industry. The industry's contribution in fertiliser diffusion in low demand areas needs to be seen in this perspective.

The interaction among the four processes creates an environment in which actual use at the micro level takes place. If the potential for fertiliser use is created by local varieties under unirrigated conditions, the adoption and diffusion of fertilisers shall be slow. Desai and Singh (1973) argued that one needs to distinguish between the two decisions that are crucial in determining the amount of fertiliser used by individuals: whether to use fertilisers and how much to use. As decision to use fertiliser is crop based, it would be desirable when analysing fertiliser use over a period of time to further identify (a) continuous users, and (b) those who discontinue use. Moreover, as the decision to use fertilisers is crop-specific, the analysis needs to be crop-specific rather than farm-specific. Analytically, fertiliser users over two points of time can be categorised in four mutually exclusive categories: (i) continuous users, (ii) users who applied only during first year and discontinued use in second year, (iii) users who did not apply fertilisers during first year and used fertilisers in second year, and (iv) non-users. This approach, though decomposes use to its various sources of growth, is problematic if the pattern of use is observed for more than two years. For example, if the use pattern is observed for three years, the analysis needs to be repeated separately for (a) year 1 and year 2, (b) year 2 and year 3.

If one is observing use pattern for three points of time, the decomposition can be attempted as:

Total growth over three years =

Growth of use by continuous uses all the three years

+ Growth of use by continuous users during t_1 and t_2

– Dropout in year t_2

+ Adopters during period t_2

+ Growth of use by continuous users during t_2 and t_3

– Dropout of year t_3

+ Adopters of year t_3

The discontinuation of fertiliser use may be on the account that the farmer may be growing the crop only during one year or may be growing the crop but may not be applying fertiliser on the crop. It has been argued (Desai and Singh, 1973) that farmers do not apply fertilisers on all the crops they grow in the year they adopt fertiliser use. The usual practice is to begin with one crop and extend the use gradually to some other crops in subsequent years. It is also possible that discontinuation of fertiliser use could take place owing to poor yield response in the previous year.

There is also a need to decompose the fertiliser use pattern of continuous use in the above formulation. Change in fertiliser use of continuous users on a crop can be decomposed as follows:

$$\Delta F = F_2 - F_1$$
$$= A_2 . R_2 - A_1 . R_1$$
$$= (\Delta A + A_1)(\Delta R + R_1) - A_1 . R_1$$
$$= \Delta AR_1 + \Delta RA_1 + \Delta R\Delta A$$

where,

F_1 & F_2 = Fertiliser use in year 1 and year 2 respectively

A = Area fertilised

R = Rate of fertiliser application

$\Delta A . R_1$ = Change in fertiliser use because of change in fertilised area

$\Delta R . A_1$ = Change in fertiliser use because of change in rate of fertiliser application

$\Delta R . \Delta A$ = Change in fertiliser use because both area and fertiliser rate changed simultaneously

ΔF = Change in fertiliser use between two points of time.

For all the crops taken together, the change in fertiliser user between time t_1 and t_2 can be decomposed as:

$$\sum_{i=1}^{n} \Delta Q_i = \sum_{i=1}^{n} \Delta A_i R_i + \sum_{i=1}^{n} \Delta R_i A_i$$

where i = 1 to n crops.

The above framework helps us to identify the processes that govern adoption and diffusion of fertiliser use among relatively late adopters of fertilisers located in low potential dry areas. Elsewhere in the country, the fertiliser use in terms of coverage of farmers as well as crops has increased substantially. In these high potential areas, the farmers have stumbled upon issues like increasing prices of fertilisers and decreasing subsidy in P_2O_5 and K_2O; increasing privatisation; depleting groundwater and increased awareness towards environmental externalities; and more importantly the emerging incentives towards production enhancement. Unfortunately, the yield augmenting technologies have bypassed the low potential forest based agriculture. Consequently, the dynamics of fertiliser use in these areas needs to be understood from the view point of farmers' lack of demand for fertilisers. In this situation, many of the emerging macro issues, if not totally lose their relevance, have inconsequential implications. Southwestern tribal belt is faced with questions of survival, where livelihood processes influence nature not because of overuse of fertilisers but because of sheer lack of production incentives. Stagnation and low level of economic activities are, thus, the cause of environmental degradation. It is not maximisation of returns that guides fertiliser use in southwestern tribal belt. It is, as would be unfolding, lack of technology transfer, poor extension services, lack of market and above all failing institution credit that engenders low level use. The above conceptual framework is appropriate to investigate the empirical nature of our investigation. Using the framework developed above, we have tried in this study to understand sources of growth in fertiliser use at the micro level.

Research Questions

The following research questions are answered in the study:

(1) When fertiliser use was adopted and how was it diffused in different crops in this tribal belt?

(2) What are the different sources of growth of fertiliser use on different crops in this area?

(3) What are the characteristics of discontinuous users on different crops?

(4) What is the relative importance of rates of application and fertilised area in the growth of fertiliser use of continuous users?

(5) What is the quantum of untapped potential for (a) non-users, (b) infrequent users, and (c) continuous users?

(6) How the inefficiency in fertiliser use is governed by factors characterised as internal and external to fertiliser users?
(7) How the growth of fertiliser use can be accelerated in low fertiliser using tribal areas?

Data and Methodology

Two different sets of data were used in the study. First, discussions with functionaries of two important institutions engaged in dissemination of farm technology (the state extension network) and supply of fertilisers in Madhya Pradesh (the cooperative supply structure). The data gave us an insight into the working of the two important institutions and their facilitating role in sustaining the demand for fertilisers in low potential areas. Two districts from western tribal belt of Madhya Pradesh (Dhar and Badwani) were selected for understanding the working of these institutions. Efforts were made to understand the linkages between the district level institutions and their apex organisations with respect to (a) constraints in transfer of agricultural technology, and (b) the smooth supply of fertilisers. This information along with the findings of the farmers' survey, not only provides understanding of factors effecting adoption, diffusion, and use patterns of fertilisers on different crops but also provides understanding of effects of different patterns of fertiliser use on its demand and the role the agricultural extension, credit cooperatives, non-institutional credit and fertiliser marketing networks have in sustaining the demand for fertiliser use in this tribal area.

As the major concern of the study is to understand the demand generating forces influencing fertiliser use in the western tribal belt, the second set of data was relating to farmers' fertiliser use patterns. For the purpose, two predominant tribal districts, Badwani and Dhar, from the belt were selected for micro-level study. For a proper coverage, two blocks were selected from each district. The selection of blocks was based on high concentration of tribal population and the levels of fertiliser use. We identified high tribal population but varying fertiliser using blocks in both the districts: In Dhar district, Kuksei represents high fertiliser use (65 kg/Ac) and high tribal population block while Dahi represented low fertiliser using (41 kg/Ac) high tribal population block. In Badwani district, Sendhwa represented high fertiliser using (36 kg/Ac) and Pati, low fertiliser using (21 kg/Ac) blocks. Tribal population in all the blocks is over 82 per cent. From each selected block, three villages were selected, one each on the basis of availability of (1) low vis-à-vis high levels of irrigation, and (2) predominant crops, foodgrains vis-à-vis cotton,

Table 3.1

Sample Size in the Selected Blocks and Villages

Village	*Dhar district*		*Badwani district*		
	Kuksee[1]	*Dahi*[2]	*Sendhwa*[1]	*Pati*[2]	*Total*
	Total farmers				
Osada[3]	—	—	—	17	17
Ganthawal[5]	—	—	—	24	24
Chocki[3]	—	—	—	33	33
Aali[3]	25	—	—	—	25
Rampura[4]	23	—	—	—	23
Lohar[3]	26	—	—	—	26
Palaasi[3]	—	26	—	—	26
Jamda[3]	—	23	—	—	23
Gangpur[3]	—	28	—	—	28
Zhopali[3]	—	—	25	—	25
Dhanori[3]	—	—	25	—	25
Shahpur[3]	—	—	26	—	26
All	74	77	76	74	301

1. High fertiliser using;
2. Low fertiliser using;
3. Low irrigation foodgrains dominated;
4. Irrigation and foodgrains dominated;
5. Irrigation and commercial crops dominated.

soyabean or groundnut. In all, six villages were selected from each district. About 300 randomly selected farmers from these villages were the units of observation (Table 3.1).

From each selected village, the plan was to randomly select 25 farmers. But this number could not be maintained in each village owing to non-availability of selected farmers. Thus, from some of the villages a larger random sample was selected for in-depth study in order to have a respectable sample (Table 3.1). Structured questionnaires were canvassed to the selected farmers to understand their endowment, income, cropping pattern, year of adoption of fertilisers, cropwise adoption, reasons for adopting fertilisers and first experience of using fertilisers; fertiliser use practices for the years 2001–02, 2002–03 and 2003–04. Apart from this information, farmers' knowledge about the recommendations on fertilisers, credit and supply constraints if any, perceptions regarding optimal level of use, level of extension support available, perceptions on recent changes in input and output prices on fertiliser use, credit availability etc. were generated to explain the emerging pattern of fertiliser use.

Analysis of these data constituted the analytical understanding that has been presented in the subsequent chapters.

Chapter 4

Fertiliser Use Environment

In order to appreciate the subsequent findings on adoption and diffusion of fertiliser use in the tribal belt of Western Madhya Pradesh, these processes have to be located within the cultural constructs and the environment within which the fertiliser use has grown. The objective of this chapter is to introduce the readers to the study area, develop an understanding relating the environment within which the fertiliser use has grown in the region and examine the pattern of fertiliser use among the sample cultivators. The chapter is divided in four sections. While the first section presents the location, soil character, availability of irrigation and cropping pattern of the study districts, the second section elaborates the two important processes that govern the farmers' demand for fertilisers: the constraints in agricultural technology adoption in the region and the role the fertiliser supply and distribution system has played in augmenting farmers' demand in this area. The third section presents the characteristics of sample farmers with respect to land holdings, irrigation, cropping pattern and level of fertiliser use. The last section summarises the major findings of the chapter.

I

The Study Districts

The study districts, Badwani and Dhar, are situated in western tribal belt of Madhya Pradesh, comprising Jhabua, Dhar, Badwani, Khargaon and Khandwa. The main inhabitants of the region are tribes like Bhil, Barela, Patelia, Bhilala and non-tribal Patidars. A large part of tribal dominated area is unirrigated, highly undulating, and rocky with thin and poor quality of top soil. On the other hand, Nimar valley, inhabited by Patidars, is highly fertile and has perennial irrigation facility. The study districts fall under low rainfall region; the average annual rainfall is about 850 mm. Monsoon failure is quite frequent phenomena in the tribal dominated area; during 50 years ending 1950, there were 14 years when rainfall was only 20 per cent of normal. In all the five years between 1999–00 and 2003–04, the area witnessed severe scarcity situation.

Soil and Irrigation

The soil of the districts can loosely be classified as (a) black cotton soil, (b) alluvial soil, and (c) shallow and rocky soil. The fertile black and alluvial soils are found on both sides of river Narmada and some patches of banks of its tributaries. These two soils are deep, fertile and capable of taking multiple crops during a year. By and large, non-tribal Patidars are located in this part of the study districts. But, a large part of both the study districts comprises of shallow, stony and poor quality soil that are capable of supporting cultivation only during kharif season. This is the heartland of tribal domination and our sample villages fall in this type of area.

Situation with respect to irrigation, in the study districts is not too happy; the main source of irrigation is through privately-owned wells, both energised and otherwise (Table 4.1). In tribal-dominated areas, nearly all irrigation is done by wells. If located in undulating topography, only rains recharge the wells. During bad monsoon years, as was the case in early 2000s, wells are not capable to irrigate beyond August. Consequently, in a large part of the tribal-dominated areas only kharif cultivation is possible. In terms of effective irrigation during rabi, evidences show that irrigated area is 15 to 20 per cent of the gross cropped area.

Table 4.1
Net Cropped Area and its Sources in Study Districts

	Badwani	*Dhar*
Percentage of cropped area irrigated	33	49
Irrigation by	**Percentage**	
Canal	10	3
Pump Set	22	44
Well	53	38
Others	15	15

Land Classification and Cropping Pattern

About 62 per cent of geographical area of the selected district is under cultivation, though depleted or otherwise forest occupies about 6 per cent and 15 per cent of the geographical area of Badwani and Dhar respectively. Dhar has about half of its cultivated area under irrigation in comparison to only a third in Badwani. This difference is reflected in both cropping pattern and cropping intensity of the study villages (Table 4.2).

As the principal agricultural season in the study districts is kharif cereals like jowar, maize, bajra, pulses like tur and mong and commercial crops like cotton and soyabean dominate the cropping pattern. With irrigation, apart from the above crops, cultivators also grow rabi wheat and gram. While rabi cultivation occupies only 14 per cent of gross cropped area in Badwani, in Dhar it is over 27 per cent. In Badwani, the cropping pattern is dominated by kharif crops like jowar,

Table 4.2
Cropping Pattern and Cropped Area in Study Districts

	Badwani	*Dhar*
Forest area as percentage of geographical area	6	15
Net cut area as percentage of geographical area	63	61
Cropping intensity (%)	118	139
Cropping pattern	**Percentage**	
Jowar	20	4
Maize	11	11
Wheat	11	19
Other cereals	5	2
Tur	2	1
Urad	4	2
Gram	2	8
Other pulses	4	3
Sugarcane	1	@
Vegetables	1	@
Chilli	2	1
Soyabean	7	30
Groundnut	9	1
Cotton	16	10
Others	5	8
Gross cropped area	100 (276016)	100 (701222)

Figure in parentheses is gross cropped area in acres.

cotton, maize, groundnut and soyabean, while in Dhar district, the cropping pattern is dominated by soyabean, wheat, maize and cotton. Over the last ten years, the area of soyabean has registered a sizeable increase replacing some area of crops like maize, jowar and kharif pulses.

II

Fertiliser Use Environment

The western tribal belt is characterised by extensive nitrogen deficiency while the levels of phosphate and potash are high in the soil. Consumption of nitrogen, therefore, is an agronomic necessity. This fact significantly influences the fertiliser use which is dominated by nitrogenous fertiliser, urea. This tribal belt consumes about 17 per cent of state's fertilisers but over 70 per cent of consumption is during kharif season. Per hectare consumption in the study districts, Dhar and Badwani, at about 46 kg, is 1.2 times higher than the state average. But high consumption is concentrated in irrigated Nimar valley that is inhabited by non-tribal, Patidars. In remote tribal areas like our study villages, the consumption is about a fourth of this. This section argues that unlike rest of Madhya Pradesh, supply push by the distribution system is the main source of growth in fertiliser use. On the other hand, intensity of irrigation in the tribal belt is low and spread of HYVs is extremely poor. In fact, the available seed-fertiliser technology has bypassed the belt.

Constraints in Agricultural Technology Transfer

After independence, three important processes have brought considerable change in constitutive ideas and behavioural expectations of tribals of Western Madhya Pradesh. Development interventions of state, political decentralisation and seasonal migration have influenced the tribal social and economic outlook. Development for tribals brings images of large projects, displacement and associated hardships. Otherwise, large sections of the Nimar tribals have missed the fruits of post-independence development interventions, which have been poor both in terms of coverage and quality. Except for elementary education and public distribution system (PDS), other amenities like access to electricity, all-weather roads, health services, drinking water and sanitation, irrigation, agricultural extension, rural cooperatives etc. are poor. In fact, their remoteness to administration and civil society institution is overriding enough to allow these facilities to reach the area. What has, however, reached them is the fear of displacement due to development projects with administrative insensitivity that accompanies them. Presence of administration—in the form of police, forest and revenue officials—is synonymous to corruption, repression and exploitation. The experiences of development in tribal consciousness imprint mistrust towards non-tribals, caution and apathy towards state intervention and loss of land and natural resources.

Colonial marginalisation of tribal interests in the Nimar valley has been discussed in the available literature (Sah and Bhatt, 2004). After independence, the ways of administrative set-up did not change, though the rational of development did. Independent India inherited an economy pervasive of intense poverty. Agricultural production was stagnating and productivity was falling. Industrial sector was small and its traditional manufacturing and trade was stunted. Moreover, the trade and infrastructure were designed to feed its colonial interests. Modern industry was discouraged and the regime flooded the Indian market with low cost industrial goods that further depressed indigenous industry. The capability of the industry to absorb the workforce was marginal. Reduced death rates and increased pressure from growing population fragmented the land and landless class increased rapidly. The system of collecting agricultural tax reinforced these tendencies. Consequently, agriculture—the dominant sector of country—was characterised by a large labour force tilling small and fragmented plots with seed varieties just capable to feed its growing population.

This scenario was responsible for a desire and hence choice of development strategy in India. Few pressing ideologies also influenced the choice. First, false optimism with respect to capabilities of agriculture originated due to good rainfall, and achievement of low targets fixed during 1951–56; second, structural constraints of agriculture were wrongly identified as *slack* between average and higher levels of yields that some farmers have achieved. It was wrongly presumed that the agricultural technology existed in the country and the *slack* was due to inadequacy of supplies and the manner of its usage; and lastly, was the tendency to equate modernisation with industrialisation.

The debate over the development strategy after independence was guided by these considerations. The prime need before the country was to accelerate the rate of material capital formation. The limiting factor in this regard was low savings. It was argued that shortage of savings was mirrored in the inadequacy of production of capital goods. And thus came forward the theory behind Mahalanobis plan that considered source of growth was capital goods and the means of growth was allocation of resources for their acquisition. The more resources so allocated, the faster production capacity would grow. The underlying assumption was that with a faster growth in production, benefits would trickle down to the masses and thereby ensuring distributive justice. But, the growth remained too little to trickle down.

The situation further deteriorated with respect to distributive justice when second generation problems of new agricultural technology started emerging in the mid-1970s. The new agricultural strategy created islands of prosperity amidst mass of poverty. Inequality across groups of people and across locations got intensified. Tribals, marginal farmers and landless labourers were losers and so were the areas that remained outside the purview of new agricultural technology. The lopsided investment in research and development between irrigated and dry farming technology affected the tribals adversely who depended on dry farming. This capital-intensive approach resulted in inequality and concentration on the one hand and insensitivity and violent social disruption on the other. Alienation, deprivation and inequality, it can be argued, are inherent in the strategy of Indian development.

The change in the form of technologically-driven agriculture, though restricted to a small subset of area, is not entirely an outcome of state intervention either. It happened because of seasonal migration. Seasonal migration for tribals is not coping with scarcity alone it is experiencing development that has bypassed them. What a migrating tribal learns during his stay as a labourer either on irrigated farms of Nimar plains or on the farms of irrigated HB cotton growers of Khandesh or on the fields of irrigated paddy and sugarcane growers in south Gujarat is a main source of agriculture technology transfer. How to energise the well and distribute water; how to grow irrigated fertiliser-responsive cotton, soyabean and wheat; utilisation of post-harvest technology of these market-oriented crops etc. is learnt during seasonal migration. No matter how much the Adivasi Mukti Sangathan and Narmada Bachao Andolan discredits it, this process of production of knowledge is creating space for non-tribal institutions that is inducing conversion of subsistence way of agriculture to market-driven agriculture (see Sah and Bhatt, 2004, for details). The process is, however, internalised without any state support. In irrigated areas, the bania is providing agricultural credit without any security. The tribals are also able to repay the loans promptly at least during good agricultural years. But, in the process, Tribals have started attaching a high degree of dependence on non-tribal institution they loathed the most, the market. This is an ideological contradiction that is the handiwork of modernity. The above description reveals that exploitation of this difficult untapped potential is not sufficient for rapid growth in fertiliser use. Equally important is the role of fertiliser supply and distribution system in this area.

Role of Fertiliser Supply System

The Ministry of Agriculture, Government of India, determines the total fertiliser supply for Madhya Pradesh, like elsewhere, under the supply plan it develops for kharif and rabi seasons. The supply plan fixes up the domestic fertiliser producers that would supply fertilisers in the state and allocate the supplementing proportion of imported fertilisers to the state. The supply plan, under retention pricing scheme, also fixes the farmers' and producers' prices.

IFFCO is the lead supplier of fertilisers in the region. The apex-distributing agency, the State Cooperative Marketing Federation, lifts allocated fertiliser to distribute it to farmers through their distribution channels, based on the requirement of the district coordination committees. There are two different channels through which farmers get their supply of fertilisers. Although two different marketing channels exist in the region but it is the cooperative distribution channel that is major distributor of fertilisers and the share of private channels is marginal. This is mainly because sale of urea is under monopoly control of cooperatives. Cooperative societies account for over 85 per cent supply of fertilisers, as government of Madhya Pradesh restricted the sale of urea by private marketing channels in 1994. There are over 440 sales points of fertilisers in Dhar and Badwani districts, about 60 per cent are cooperative points that are located in the villages. The private sales points selling fertilisers other than urea are located at taluka headquarters.

The strengths of cooperative channel are: (i) inexpensive marketing credit provided to village cooperatives by district cooperative banks, (ii) apex marketing agency's attention to placement logistics rather than inventory carrying cost, (iii) coordination between different agencies, especially between fertiliser producers, village society and district marketing federation, and (iv) advance lifting and storage by apex marketing agency that works as a supply push to increase the sales.

Even if there is no special incentive as marketing margins and subsidised carrying cost, the central government's policy to subsidise warehouse construction has resulted in large storage space in tribal areas that is being hired by fertiliser producers.

This makes an economic sense for, marketing federation's overall economy improves because of storing fertilisers for producers. This also results in smooth movement of fertilisers, and supply pressures on cooperatives to sell fertilisers. Marketing federation passes the marketing margins to selling points, as given by the manufacturers (Table 4.3). The amount differs for different fertilisers but in the case of urea it is Rs 200 per tonne, of which Rs 20 is retained by the apex body and Rs 180 is passed to cooperative societies for lifting the fertilisers from rail head, its storage, carrying cost and other incidental charges that includes society's profit.

Table 4.3
Marketing Margins and Selling Price of Fertilisers

Quality	*Rs per tonne*			
	Cost of fertiliser	*Cooperative* margin passed on by MARFED*	*Commission of MARFED*	*Margin passed by fertiliser manufacturer*
Urea	5060	180	20	200
Single Super Phosphate	3140	118	42	160
DAP	9800	253	47	300
12:32:16	8800	253	47	300
Potash	4660	148	23	170

*Also transfer the 9.2 per cent of total margin for paying VAT.

In absence of a substantial growth in irrigation in this tribal belt and limited scope of diffusion of HYVs, the supply side has become an important force in generating growth in fertiliser use. The important strength of the cooperative distribution channel is its presence at the village level. Facilities of cheap marketing credit to cooperatives and production credit to farmers for purchase of fertilisers through cooperative societies is an aggressive marketing strategy in the cooperative supply line. Credit limits fixed by district cooperative banks help movement of fertilisers from apex to village sales points in the cooperative marketing structure. But a low coverage of working credit to farmers owing to default, lack of land records and at times indifferent attitude of officials of cooperatives for creating additional demand are responsible, at present, for system's inability to vigorously push fertilisers. There is a need to change the attitude of cooperatives if growth of fertiliser use has to increase in this region. The urea prices have been quite stable but output prices have increased significantly. Price of marketed crops specially soyabean increased from Rs 1,000 to Rs 1,400, cotton from Rs 1,800 to Rs 2,200 and groundnut from Rs 1,400 to Rs 1,800 in three years ending 2003–04.

To sum up, the environment, thus, is dominated by supply push but demand-generating forces are relatively inactive in this tribal region.

III

Sample Cultivators

Size of Operational Holding and Tenancy

Middle size farmers dominate the cultivators in the sample area; the average size of holding of the sample farmer is about 4.4 acre. Almost 24 per cent cultivators operate less than 1.5 acre land, 38 per cent operate land holdings ranging between 1.5 acres and 3.5 acres and 38 per cent cultivators operate over 3.5 acre land (Table 4.4). Small farmers account for only 7 per cent of total cultivated area in the sample, medium size farmers control about 22 per cent and large size cultivators account for as high as 71 per cent of total cultivated area. About 6 per cent of the sample farmers (18 in all) have reported short-term land transaction in the form of land leasing in and out. While large size farmers have reported leasing land out, all size class of farmers has reported leasing in.

Table 4.4
Tenancy Status under Different Size Class of Holdings

Area leased (Acre)	*number of farmers*			*All*
	Small farm	*Medium farm*	*Large farm*	
–5	—	—	1	1
–2	—	—	1	1
–1	—	—	1	1
None	67	108	108	283
0.5	1	—	—	1
1	3	2	2	7
1.5	1	—	—	1
2.5	—	1	—	1
3	—	2	—	2
4	—	—	1	1
5	—	—	1	1
6	—	—	1	1
Total	72 (24)[a] [7][b]	113 (37) [21]	116 (39) [71]	301 (100) [100]

a Percentage of total farmers
b Percentage of total cropped area

Cropping Pattern and Irrigation

Maize, jowar, cotton and bajra are four main crops grown by a large number of farmers (Table 4.5). However, the cropping pattern varies across different size class of farms. While jowar, cotton and wheat are three important crops on small farms, in the case of medium size farms jowar, maize and bajra are important and in large farms cotton, jowar and maize are three important crops. On the other hand, commercial crops, soyabean and cotton, are relatively more important on large farms.

Table 4.5
Cropping Pattern under Different Size Class of Holdings, 2002–03

Crops	*Upto 1.50 acre*	*1.51 to 3.50 acre*	*> 3.50 acre*	*All*
Maize	21.4 (43)[a]	22.5 (74)	16.4 (88)	18.1 (205)
Soyabean	8.7 (14)	6.2 (17)	10.6 (41)	9.5 (72)
Jowar	21.9 (34)	23.3 (68)	18.4 (85)	19.7 (187)
Cotton	18.4 (24)	15.8 (49)	20.2 (68)	19.1 (141)
Kharif Pulses	1.7 (4)	4 (19)	3.5 (29)	3.4 (52)
Groundnut	3.7 (6)	4.6 (16)	6.5 (33)	5.9 (55)
Wheat	12.6 (21)	6.2 (16)	9.8 (48)	9.2 (85)
Gram	—	—	1.4 (4)	1 (4)
Chilli	1.1 (3)	0.2 (1)	5.1 (25)	3.8 (29)
Bajra	10.6 (20)	17.2 (43)	8.2 (45)	10.3 (108)
All crops	100 [103.41][b]	100 [298.5]	100 [992.57]	100 [1394.38]

a Figures in the parnetheses are number of farmers.
b Figures in the parentheses show the actual area in acres.

About 26 per cent of gross cropped area of the sample farmers has irrigation facilities (Table 4.6). Surprisingly, over two-thirds of total irrigation is used for rabi season, specifically wheat cultivation. Over 96 per cent of wheat acreage and 82 per cent of wheat growers irrigate their wheat. Medium size farms have relatively larger proportion of gross cropped area under irrigation compared to small and large farms. Irrigation is highly concentrated on a few crops; 97 per cent wheat farmers irrigate it, over 40 per cent of farmers who have grown chilli and cotton have irrigated the crop, over a third of the cultivators who have grown maize have irrigated the crop. In terms of the cropped area irrigated, 91 per cent wheat area receives irrigation, 59 per cent chilli area and 45 per cent of cotton area receives irrigation.

Table 4.6
Percentage Irrigated Area under Different Crops, 2002–03

Crops	*Upto 1.50 acre*	*1.51 to 3.50 acre*	*> 3.50*	*All*
Maize	19.9	11.9	17.9	16.5 (18)[a]
Soyabean	5.6	12.4	14.3	13.4 (14)
Jowar	6.6	8.6	7.7	7.9 (1)
Cotton	40.4	21.3	51.1	45 (41)
Kharif Pulses	—	8.3	—	2.1 (2)
Groundnut	—	—	3.8	3 (5)
Wheat	100	100	88.1	91.1 (96)
Gram	—	—	22.2	22.2 (25)
Chilli	100	—	58.6	58.9 (45)
Bajra	—	1.9	—	0.7 (1)
All crops	27.3 [28.19][b]	15.7 [46.8]	28.3 [281.25]	25.6 [356.34]

a Figures in the parnetheses are number of farmers.
b Figures in the parentheses show the irrigated area.

Fertiliser Use

Out of 301 sample farmers, 262 have applied fertilisers during 2001–02, 268 used in 2002–03 and 270 farmers used fertilisers in 2003–04 (Chart 4.1). About 18 farmers had never used fertilisers, while seven farmers after adopting fertilisers discontinued its use for last three years. It means that out of 301 farmers, 39 did not apply fertilisers in first year, 33 and 31 farmers respectively did not apply fertilisers in second and third year. Out of total 262 farmers who had applied fertilisers in 2001–02, 255 applied fertilisers continuously all the three years. Interestingly, there are seven farmers who adopted fertilisers as late as 2001–02. It is also worth noting that fertiliser users over the years have increased in the sample.

Chart 4.1

Pattern of Fertiliser Use over Years Use

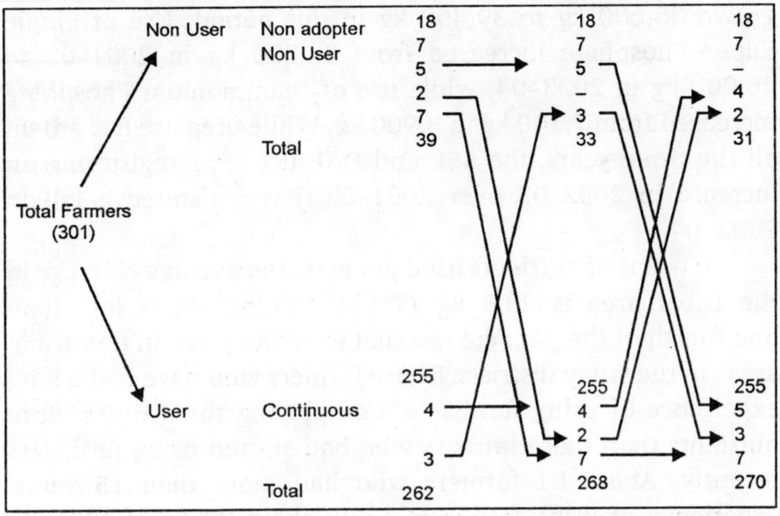

Nearly all the fertiliser users applied urea on their crops but Single Super Phosphate (SSP) and Diammonium Phosphate (DAP) users were only about 45 and 24 per cent of total fertiliser users (Table 4.6). In terms of total use during 2001–02, urea consumption was 35,600 kg, followed by SSP at

Table 4.6

Fertiliser Use Pattern of Sample Farmer

	2001–02	*2002–03*	*2003–04*
		Number of farmers	
Non-user	7	7	7
Non-adopters	18	18	18
Urea	259 (35605)*	267 (38957)	267 (39382)
DAP	63 (7147)	72 (9500)	65 (8925)
SSP	118 (24190)	115 (27358)	124 (26871)

* Figure in the parentheses are total use of the material.

24, 200 kg and DAP consumption was 7,100 kg. In three years between 2001–02 and 2003–04, fertiliser use as well as all the material used by the sample farmers has grown. Urea use has grown 35,600 kg to 39,400 kg in this period. Use of Single Super Phosphate increased from 24,200 kg in 2001–02 to 26,900 kg in 2003–04, while use of Diammonium Phosphate increased from 7,100 kg to 8,900 kg. While urea use has grown all the three years, the SSP and DAP use after registering an increase in 2002–03 over 2001–02, has registered a fall in 2003–04.

In terms of nutrients used per acre, the average NPK use in the tribal area is 23.8 kg (Table 4.7), which is less than one-fourth of the per acre use that is taking place in non-tribal areas in the study districts. Those farmers who have had a long experience of using fertilisers are applying three times more nutrients than those farmers who had started using fertilisers recently. About 81 farmers who had more than 15 years' experience of using fertilisers apply about 33 kg of NPK per acre while those who had started using fertilisers in last five

Table 4.7
Kilogram Per Acre Nutrient Used by Sample Farmers, 2002–03

Duration of use	*Number of farmers*	*Net cultivated area (acre)*	*NPK (kg/acre)*
Less than 5 years	63	184.7	12.0
5 to 15 years	124	519.4	18.8
15 years or more	81	490.5	33.4
Size class of holding			
Small	60	71.5	36.1
Medium	101	252.5	17.9
Large	107	870.6	24.4
All Farmers	268	1194.6	23.8

years have applied only 12 kg of NPK per acre. The application of fertilisers by different size class of farmers reveals that small farmers apply relatively more nutrients per unit of area than the other farmers. In a sense, this indicates that small farmers are early adopters and medium size farmers are late adopters of fertilisers.

The above discussion raises issues relating to adoption and diffusion of fertilisers by sample cultivators. It also raises questions relating to source of growth and crops that are dominating in this process. The subsequent chapters analyse these processes.

IV

Summing up

The inhabitants of the region are tribes like Bhil, Barela, Patelia, Bhilala and non-tribals like Patidars. A large part of tribal-dominated area is unirrigated, highly undulating, and rocky with thin and poor quality of top soil. On the other hand, Nimar valley, inhabited by Patidars, is highly fertile and has perennial irrigation facility. The study area also falls under low rainfall region and monsoon failure is a quite frequent phenomena here. Soils are capable to support cultivation only during kharif season. Western tribal belt is also characterised by extensive nitrogen deficiency while the levels of phosphate and potash are high in the soil. Consumption of nitrogen, therefore, is an agronomic necessity. This fact significantly influences the fertiliser use which is dominated by nitrogenous fertiliser, urea. Although the per hectare consumption in the study districts is higher than the state average, the high consumption is concentrated in irrigated Nimar valley that is inhabited by non-tribal Patidars. In remote tribal areas like our study villages, the consumption is about a fourth of this.

Unlike irrigated Madhya Pradesh, supply push by the distribution system is the main source of growth in fertiliser use. On the other hand, intensity of irrigation in the tribal belt is low and spread of HYVs is extremely poor. In fact, the available seed-fertiliser technology has bypassed the belt. The change in the form of technologically-driven agriculture, though restricted to a minuscule area, is not entirely an outcome of state intervention either. It happened because of seasonal migration. Seasonal migration for tribals is not coping with scarcity alone it is experiencing development that has bypassed them.

IFFCO is the lead fertiliser supplier of the region. The apex-distributing agency, the State Cooperative Marketing Federation lifts allocated fertilisers to distribute it to farmers through their distribution channels, based on the requirement of the district coordination committees. Although two different marketing channels exist in the region but it is the cooperative distribution channel that is major distributor of fertilisers (85%) and the share of private channels is marginal (15%); mainly because the sale of urea is under monopoly control of cooperatives. The private sales points selling fertilisers other than urea are located at taluka headquarters. The strengths of cooperative channel are: (i) inexpensive marketing credit provided to village cooperatives by district cooperative banks, (ii) apex marketing agency's attention to placement logistics rather than inventory carrying cost, (iii) coordination between different agencies, especially between fertiliser producers, village society and district marketing federation, and (iv) advance lifting and storage by apex marketing agency that works as a supply push to increase the sales.

The other strength of the cooperative distribution channel is its presence at the village level. Facilities of cheap marketing credit to cooperatives and production credit to farmers for purchase of fertilisers through cooperative societies is aggressive marketing strategy in the cooperative supply line. Credit limits fixed by district cooperative banks help movement of fertilisers from apex to village sales-points in the cooperative marketing structure. But a low coverage of working credit to farmers owing to default, lack of land records and at times indifferent attitude of officials of cooperatives for creating additional demand are responsible, at present, for system's inability to vigorously push fertilisers. In absence of a substantial growth in irrigation in this tribal belt and limited scope of diffusion of HYVs, the supply side has become an important force in generating growth in fertiliser use.

Out of 301 sample farmers, 262 have applied fertilisers during 2001–02, 268 used in 2002–03 and 270 farmers used fertilisers in 2003–04. About 18 farmers had never used fertilisers, while 7 farmers after adopting fertilisers discontinued its use for last three years. It means that out of 301 farmers, 39 did not apply fertilisers in first year, 33 and 31 farmers respectively did not apply fertilisers in second and third year. Out of total 262 farmers who had applied fertilisers in 2001–02, 255 applied fertilisers continuously all the three years. Interestingly, there are seven farmers who adopted fertilisers as late as 2001–02. In terms of nutrients use per acre, the average NPK use in the tribal area is 23.8 kg, which is less than one-fourth of the per acre use that is taking place in non-tribal areas in the study districts. Those farmers who have had a long experience of using fertilisers are applying three times more nutrients than those farmers who had started using fertilisers recently. The application of fertilisers by different size class of farmers reveal that small farmers apply relatively more nutrients per unit of area than the other farmers. In a sense, this indicates that small farmers are early adopters and medium size farmers are late adopters of fertilisers.

of fertilisers from apex to village sales points in the cooperative marketing structure. But a low coverage of working credit to farmers owing to default, lack of land records and at times indifferent attitude of officials of cooperatives for creating additional demand are responsible, at present for system's inability to vigorously push fertilisers. In absence of a substantial growth in irrigation in this tribal belt and limited scope of diffusion of HYVs, the supply side has become an important force in generating growth in fertiliser use.

Out of 301 sample farmers 262 have applied fertilisers during 2001–02, 268 used in 2002–03 and 270 farmers used fertilisers in 2003–04. About 18 farmers had never used fertilisers, while 7 farmers after adopting fertilisers discontinued its use for last three years. It means that out of 301 farmers, 39 did not apply fertilisers in first year, 33 and 31 farmers respectively did not apply fertilisers in second and third year. Out of total 262 farmers who had applied fertilisers in 2001-02, 255 applied fertilisers continuously all the three years. Interestingly there are seven farmers who adopted fertilisers as late as 2001-02. In terms of nutrients use per acre, the average NPK use in the tribal area is 23.4 kg, which is less than one-fourth of the per acre use that is taking place in non-tribal areas of the study districts. Those farmers who have had a long experience of using fertilisers are applying three times more nutrients than those farmers who had started using fertilisers recently. The application of fertilisers by different size class of farmers reveal that small farmers apply relatively more nutrients per unit of area than the other farmers. In a sense, this indicates that small farmers are early adopters and medium size farmers are late adopters of fertilisers.

Chapter 5

Adoption and Diffusion of Fertilisers

The main objectives of this chapter are to: (a) analyse the time dimension of adoption of fertilisers and quantity of fertilisers used by first time users in the sample for different crops, (b) understand why farmers adopted fertilisers and their experiences after first use, (c) understand the diffusion of fertiliser use to other crops and time lag in the diffusion process, (d) estimate the growth in fertiliser use owing to diffusion process, and (e) understand factors behind delay in adoption of fertiliser by some cultivators.

Adoption of Fertilisers

Desai (1969) argued that fertiliser use in appreciable quantities began in India in the 1920s when tea plantations started using imported ammonium sulphate. Outside plantation, fertiliser use started in 1930s when it was extended to sugarcane and rice. The Government of India policy helped fertiliser use spread to foodgrains, owing to Grow More Food campaign in early 1940s. On the eve of independence, there was a complete transformation of the environment whereby fertilisers were recognised as surest and the quickest means of increasing foodgrain production. The use during pre-HYVs period, however, concentrated on crops like irrigated rice, wheat, sugarcane, potato, cotton, tobacco and banana. Desai, Chary and Bandyopadhyay (1973) analysing the adoption behaviour at micro level in an area dominated by commercial crop grown under irrigated conditions, find that the fertiliser use began in 1940. But, the process remained slow for next 20 years. The most rapid diffusion of fertiliser took place between 1960 and 1966. They also find that early adopters fertilised only one crop in the year of adoption. Most cultivators fertilised paddy, tobacco and chilli. Even after ten years, a significant proportion of users did not adopt fertilisers on jowar, bajra and pulses.

Unlike a highly irrigated area, in our sample of tribal farmers, adoption of fertiliser use started as late as 1970. As the recall of the year of fertiliser adoption for a long span of time is problematic, at the best the year should be indicative of the adoption process rather than a precise number. The important crops fertilised by early adopters were maize, groundnut, cotton and wheat (Table 5.1). For the first 10 years, only a few farmers adopted fertiliser use. It was only between 1992 and 2003 that there was sizable increase in fertiliser adoption; it took over 20 years for first 50 per cent of farmers to adopt fertilisers, while in next 10 years rest 50 per cent farmers adopted fertilisers. Still, even in 2003, 6 per cent of sample farmers (18 in all) did not adopt fertilisers. Bulk of these farmers are illiterate having total income of household less than Rs 10,000 per annum.

Out of 283 sample cultivators who had adopted fertilisers, over 57 per cent adopted on maize. The next important crop on which farmers adopted fertilisers is jowar (20% of adopters) followed by cotton (9%) and wheat (7%). None of the sample cultivators adopted fertiliser use with pulses (both kharif and rabi). On the other hand, very few (less than 5) farmers

Table 5.1
Year of First Fertiliser Use and the Fertilised Crop

Fertiliser use beginning	*Maize*	*Soyabean*	*Jowar*	*Cotton*	*Groundnut*	*Wheat*	*Chillies*	*Bajra*	*Total*	*%*	*CUM %*
				Number of Farmers Adopting Fertilisers on a Crop							
1970	1	—	—	—	1*	—	—	—	2	0.7	0.7
1978	2	—	—	1	—	—	—	—	3	1.1	1.8
1980	9	—	—	2	—	2	—	—	13	4.6	6.4
1981	2	—	—	—	—	—	—	—	2	0.7	7.1
1982	1	—	—	—	—	—	—	—	1	0.4	7.5
1983	7	—	—	3	—	1	1	—	12	4.2	11.7
1984	2	—	—	—	—	—	—	—	2	0.7	12.4
1985	14	—	—	—	—	1	—	—	15	5.3	17.7
1986	5	—	—	—	—	—	—	—	5	1.8	19.5
1987	7	—	1	—	—	1	—	—	9	3.1	22.6
1988	13	—	2	2	—	2	—	—	19	6.7	29.3
1989	1	—	—	—	—	—	—	—	1	0.4	29.7
1990	21	—	11	3	1	1	—	1	38	13.4	43.1
1991	4	—	3	—	—	—	—	—	7	2.5	45.6

Cont'd...

Cont'd...

1992	9	—	2	2	1	—	—	1	15	5.3	50.9
1993	10	—	14	1	1	2	—	—	28	9.9	60.8
1994	3	—	4	2	—	—	—	1	10	3.5	64.3
1995	4	—	3	1	—	2	—	—	10	3.5	67.8
1996	2	—	2	1	—	1	—	—	6	2.1	69.9
1997	6	1	3	1	—	5	—	1	17	6.0	75.9
1998	16	—	5	1	—	—	—	2	24	8.5	84.4
1999	9	1	2	1	—	—	—	1	14	4.9	89.3
2000	9	—	3	—	1	1	—	2	16	5.7	95.0
2001	5	—	1	1	—	—	—	—	7	2.5	97.5
2002	1	—	1	—	—	—	—	—	2	0.7	98.2
2003	—	—	1	3	—	—	—	1	5	1.8	100
Total	163	2	58	25	5	19	1	10	283 **	100	

*Information relating to diffusion of fertilisers is not available from the farmer due to memory loss.

**Out of 270 fertiliser users in 2003, 7 farmers did not use fertilisers between 2001 and 2003, and 6 dropped out in 2003.

Source: Based on primary data collected by the author.

adopted fertiliser use with groundnut, soyabean and chillies. While early adopters started fertiliser use with maize, cotton, wheat, groundnut and chilli, the late adopters started fertilising their crops with soyabean and bajra.

Interestingly, bulk of early fertiliser adopters on maize were unirrigated farmers. A sizable area under irrigation in crops that were fertilised was in the case of soyabean, wheat and cotton (Table 5.2). For rest of the crops, less than 7.8 per cent of the crop area under fertilisers was irrigated. In fact, when fertiliser use was adopted first time on groundnut, chillies and bajra, it was adopted on totally unirrigated area. Urea was the main fertiliser used by the farmers when fertilisers were first adopted, followed by Single Super Phosphate (SSP) and Diammonium Phosphate (DAP). None of the farmers used potash when applied fertilisers first time. In terms

Table 5.2

Crop Specific Use of Fertilisers when Used First Time

Crops	*No. of farmers*	*Cultivated area (Ac)*	*Irrigated area (Ac)*	*Fertilised area (Ac)*	*DAP (kg)*	*SSP (kg)*	*Urea (kg)*	*NPK (kg)*
Maize	163	291.3	22.6	229.2	203	595	6784	3346 (14.6)
Soyabean	2	2.0	1.0	1.0	25	—	—	16 (16.0)
Jowar	58	98	1.5	85.3	100	75	1855	929 (10.9)
Cotton	25	51.7	21.9	51.7	960	1598	1841	1749 (33.8)
Groundnut	5	4	—	4	—	100	100	62 (15.5)
Wheat	19	44.8	20	40.7	30	525	985	556 (13.7)
Chilli	1	2	—	2	—	—	50	23 (11.5)
Bajra	10	17.2	0	15.1	—	—	210	97 (06.4)

Source: Based on primary data collectded by the author.
Figures in parentheses are NPK in kg per acre.

of nutrient use per unit of area, highest use by first time adopters was on cotton (33.8 kg per fertilised acre). The intensity of fertiliser use on maize, the crop on which maximum number of farmers adopted fertilisers, was only 14.6 kg per acre. Surprisingly, fertiliser use on maige two other irrigated crops soyabean and wheat by first time adopters was only 16 kg per acre and 13.7 kg per acre respectively. In fact, except for cotton, there is no significant variation in fertiliser use per unit of area of first time users. This indicates that the farmers used fertiliser very cautiously when they fertilised their crop for the first time; use was confined only to one crop and at reasonably low rates, with only exception of cotton.

Those who adopted fertilisers have given a number of reasons for doing so. For increasing production; after seeing others fertilise their crops and for experimenting are the three most important reasons for adopting fertilisers. Although, fertilisers were cheap and freely available (in the form of experiment kits supplied by extension agency), and credit was readily available were reasons given by a minuscule of cultivators, it is the pressure to increase yield that motivated farmers to adopt fertiliser use (Table 5.3). Interestingly, about a tenth of cultivators adopted chemical fertilisers because of unavailability of farmyard manure (FYM). This, in a sense, reflects on the awareness about nutrient deficiency of cultivators located in remote rural areas, under severely difficult farming conditions and devoid of a viable agricultural technology. Availability of farmyard manure in this tribal belt is severely constrained owing to low cattle population and lack of stall-feeding. As a result, adoption and diffusion of fertiliser use in this area is also a soil management device under low availability of FYM. But relatively late adoption of fertilisers in the region is a reflection of poor soil and high risk associated with fertiliser use.

Table 5.3
Two Important Reasons for Adoption of Fertilisers

Reason	*Started 1998 to 2003*	*Started 1989 to 1997*	*Started 1988 or earlier*	*Total*
		Percentage		
After seeing others	30.9	40.9	27.7	34.6
For increasing production	45.7	74.2	59.0	62.9
For experiment	13.2	18.2	37.3	22.6
Free availability	0.0	0.8	3.6	1.4
Motivated by institutions	1.5	6.8	13.3	7.4
From friends, relatives	19.1	6.1	6.0	9.2
Big farmers	8.8	2.3	6.0	4.9
Migration	1.5	0.0	0.0	0.4
FYM non-availability	17.6	6.8	7.2	9.5
Increased irrigation	1.5	2.3	1.2	1.8
Fertilisers were cheap	1.5	3.8	6.0	3.9
I don't know	5.9	3.0	4.8	4.2
Availability of funds	8.8	2.3	0.0	3.2
For improving crop quality	1.5	4.5	3.6	3.5
Total responded	(68)	(132)	(83)	(283)

Source: Based on the primary data collected by the author.

This indicates that tribal farmers were well aware about the nutrient deficiency but the physical condition of their land—highly undulating land with no irrigation facilities and thin top soil—turned them into late adopters of fertilisers. There is no significant difference across time of adoption and the reasons for adopting fertiliser use in the sample except that the early adopters were quite fascinated by the experiences of others and wanted to emulate their success stories; the reasons like after seeing others and for experiment are the answers given by significantly large number of cultivators among those who adopted fertilisers before 1989 compared to late adopters. The late adopters were influenced much more by the advise of friends and big farmers and experienced serious nutrient deficiency owing to non-availability of FYM. The change in the form of technologically-driven agriculture, though not restricted to minuscule of irrigated area, is not entirely an outcome of state intervention. It happened because of seasonal migration. Seasonal migration for the tribals is not coping with shocks alone (Sah and Bhatt, 2004); it is, as explained earlier, experiencing development that has bypassed them.

What a tribal learns during his stay as a migrant labourer—either on irrigated farms of Nimar plains or on the farms of irrigated HB cotton growers of Khandesh or on the fields of irrigated paddy and sugarcane growers in south Gujarat—is a main source of agricultural technology transfer. How to energise the well and distribute water, how to grow irrigated fertiliser-responsive cotton, soyabean and wheat, and utilisation of post-harvest technology of these market-oriented crops, etc. are learnt during seasonal migration. No matter how much the civil society activist in the area discredits it (ibid.), this process of production of knowledge has created space for non-tribal institutions, that is, inducing conversion of subsistence way of agriculture to a market-driven agriculture. The process is, however, internalised without any state support. In the process, they have started attaching a high degree of dependence on non-tribal institution they loathed the most, the market.

The experiences of first-time adopters of fertiliser use have been quite encouraging but not without dissent (Table 5.4). Positive feelings towards increasing fertiliser use and diffusing it to unfertilised areas is governed by these experiences. Increased production due to fertiliser use, improvement in quality of production and crop maturing early are the most frequently reported experiences given by substantial number of the farmers. Many other positive experiences like

Table 5.4
Two Important Experiences of Farmers after Adoption of Fertilisers

	Started 1998 to 2003	*Started 1989 to 1997*	*Started earlier to 1988*	*Total*
	Percentage of farmers			
Production increase in	54.4	63.6	62.7	61.1
Early crop harvest	—	6.8	3.6	4.2
Plants mature faster	22.1	20.5	20.5	20.8
All crop production increased	5.9	11.4	10.8	9.9
Crop quality improved	33.8	34.1	36.1	34.6
Crop condition improved	8.8	21.2	15.7	16.6
Diffusion to other crops	—	4.5	—	2.1
SSP responded in FYM presence	1.5	—	—	0.4
Some crops (cotton) responded	—	2.3	4.8	2.5
Experimented on maize	4.4	0.8	1.2	1.8
Used first time for experiment	2.9	—	—	0.7
Don't know	—	1.5	2.4	1.4
Diffused in cotton if funds available	1.5	—	—	0.4
Fertility increased	1.5	2.3	0.0	1.4
Moisture decreased rapidly	1.5	0.8	1.2	1.1
Fertiliser was harmful	1.5	—	—	0.4
No significant different	7.4	3.8	7.2	5.7
Water scarcity, no response	8.8	3.8	2.4	4.6
More irrigation with fertiliser	1.5	2.3	3.6	2.5
Without fertiliser yield reduces	2.9	3.8	4.8	3.9
Crop did not respond	1.5	—	1.2	0.7
Crop growth but poor yield	4.4	—	2.4	1.8
Total respondents	(68)	(132)	(83)	(283)

yield-response of fertilisers on specific crop or fertiliser material response, or improvement in soil fertility were also reported. But about 17 per cent of farmers did report negative experiences like poor crop response, rapid loss of moisture owing to fertiliser use and harmful effects on crop. Nonetheless, positive effects of fertiliser use dominate the tribal psyche.

It would be worth understanding how the adopter who fertilised first time a crop, diffused fertilisers to different other crops over the time. Equally interesting shall be to analyse the time-lag with which such diffusion took place.

Diffusion of Fertilisers

Table 5.5 gives the pattern of diffusion of fertiliser use after adopting it on a crop by sample farmers. Although it was maize that saw maximum first adoption, the diffusion to other crops in the case of wheat and groundnut is much faster. It should,

Table 5.5

Diffusion of Fertiliser after Adopting it on a Crop

Crops	*Maize*	*Jowar*	*Cotton*	*Wheat*	*Bajra*	*Groundnut*	*Soyabean*	*Chilli*
				Number of Farmers				
Maize	(163)*	17	17	13	3	2	1	1
Jowar	51	(58)*	10	8	2	—	—	—
Cotton	99	20	(25)*	12	1	4	1	1
Wheat	56	11	11	(19)*	3	2	1	—
Bajra	9	17	3	4	(10)*	—	—	—
Groundnut	15	10	3	2	—	(5)*	—	—
Soyabean	30	4	1	2	—	2	(3)*	—
Chilli	24	1	5	4	1	2	—	(1)*
K. Pulses	10	1	2	2	—	—	—	—
Gram	5	—	1	2	1	—	—	—

*Figures in the parentheses are the number of farmers first adopted fertilisers on the crop (reported in the column head). The remaining column numbers are farmers diffusing fertilisers (on row crops) after adoption.

however, be noted that like adoption, diffusion of fertiliser use is also based on recall; hence the findings are indicative of diffusion process.

Diffusion process of different crops reveals two patterns. First, number of crops on which fertilisers were diffused significantly varies after first use on a crop. Second, diffusion on some crops like gram, kharif pulses and bajra was relatively slow, no matter with which crop the fertiliser use was first adopted. Diffusion process of 163 farmers who adopted fertiliser use on maize reveals that, 99 also started using fertiliser on cotton, 56 of them used fertilisers subsequently on wheat, 51 added fertiliser use on jowar, 30 on soyabean, 24 on chillies, 15 on groundnut, 10 on bajra, 10 on kharif pulses and five on gram. For other crops, the pattern is given in Table 5.5. After adopting fertiliser use on maize, 163 farmers diffused it to 299 other crops giving an average less than two crops (1.83) fertilised by them over time. In the case of five growers who adopted fertiliser for the first time on groundnut, diffused to another 12 crops giving an average additional 2.4 crops fertilised over time. This means that for some crops like maize a number of farmers adopted fertilisers without a time lag, for others, like groundnut and wheat, though adoption of fertilisers is slow, the diffusion of fertiliser use to other crops is relatively much faster.

This contrast in diffusion process emerges clearly when we observe that less than 16 per cent of growers who adopted fertiliser on either maize, cotton or jowar, fertilised their pulse and bajra at all. Evidently, a very large proportion of growers diffused fertiliser use on maize, cotton and wheat when they adopted fertilisers with bajra. The above pattern needs to be appreciated within the time lag of fertiliser diffusion. For all those crops on which the first adoption is by a relatively small number of farmers, like on chillies, soyabean, groundnut and bajra, revel that subsequent diffusion to other crops has been relatively rapid, within two years (Table 5.6). On the other hand, when fertilisers were adopted on maize, cotton and wheat, the subsequent diffusion to other crops took significantly more time.

Observing the adoption and diffusion of fertiliser use on three important crops maize, jowar and cotton (Appendix 5.1), it is noteworthy that (i) early adopters fertilised mainly maize, jowar, cotton and wheat, and adding over the years soyabean, bajra, gram and sometimes kharif pulses; (ii) late adopters mainly started fertiliser use with maize, cotton, soyabean, chillies, wheat and kharif pulses. As maize, jowar, cotton and wheat are dominant crops of the farmers (Table 5.5), the above pattern is not surprising. What surprises us is the late diffusion of fertilisers on cotton and soyabean. Cotton and soyabean are non-traditional crops for sample farmers for, their inclusion in cropping patterns has taken place with considerable time lag across farms. Consequently, the diffusion of fertiliser use on cotton and soyabean has also taken place with significant time lag after adoption of fertilisers on either maize or jowar.

Table 5.6

Time Lag between First Use of Fertiliser on a Crop and its Diffusion on Other Crops

First fertilised crop	*Number of farmers*	*Subsequently fertilised crop*	*Time gap (in years)*				
			Upto 2 years	*3 to 5 years*	*6 to 10 years*	*> 10 years*	
			Number of crops				
Maize	163	299	88 (29)	91 (31)	79 (26)	41 (14)	(100)
Jowar	58	81	37 (46)	27 (33)	16 (20)	1 (1)	(100)
Cotton	25	53	21 (40)	16 (30)	11 (21)	5 (9)	(100)
Wheat	19	49	22 (45)	15 (31)	5 (10)	7 (14)	(100)
Bajra	10	11	8 (73)	2 (18)		1 (9)	(100)
Groundnut	5	12	5 (42)	6 (50)	1 (8)		(100)
Soyabean	2	3	3 (100)				(100)
Chillies	1	2	2 (100)				(100)

There is another aspect of diffusion of fertiliser use. From one farmer using on one crop in 1970, fertiliser use has diffused to 283 farms (Table 5.6). On an average 3 crops (2.8) are fertilised by farmers now. From 2 acre of fertilised area, the area has increased, due to addition of new crops, to about 1148 acre, over last 23 years. If we assume that first-time adopter kept his fertiliser use on the adopted crop same, one can estimate that a total of 5,997 kg of DAP; 16,964 kg of SSP and 33,548 kg of urea has been added over 23 years since 1970. In terms of total nutrient use, this is about 21,952 kg of NPK. This is an indifferent growth in fertiliser use (Chart 5.1), for in terms of NPK use per acre, it is less than 20 kg.

Chart 5.1
Cumulative Fertiliser Use

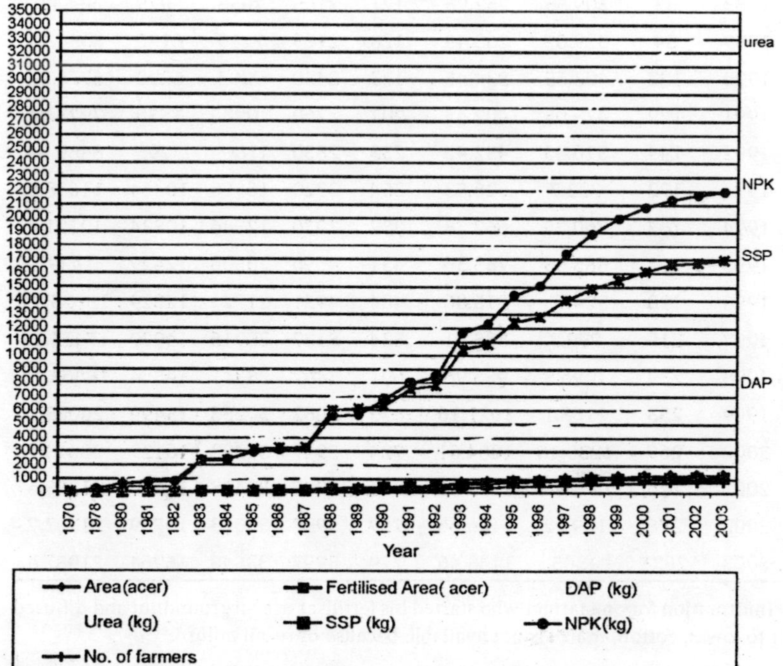

Table 5.7
Cumulative Fertiliser Use due to Diffusion of Fertilisers

Year	*No. of farmers*	*Area (Ac)*	*Fertilised area (Ac)*	*No. of crops*	*DAP use (kg)*	*Urea use (kg)*	*SSP use (kg)*	*NPK use (kg)*
1970	2*	10	2	1	0	50	0	23.0
1978	5	13.5	5.5	4	150	400	20	283.2
1980	18	42.5	19.5	11	300	785	420	620.3
1981	20	57.5	30.5	14	300	1085	420	758.3
1982	21	63.5	34.5	16	300	1185	420	804.3
1983	33	130	92.5	36	1000	3055	2320	2384.5
1984	35	133	95.5	38	1000	3135	2320	2421.3
1985	50	156.4	117.8	51	1125	3810	3145	2943.8
1986	55	171.4	125.3	58	1155	4085	3245	3105.5
1987	64	187.9	141.8	65	1180	4350	3295	3251.4
1988	83	309.55	252.95	121	2165	7055	5975	5554.9
1989	84	329.05	263.45	128	2165	7219	6125	5654.34
1990	122	403.55	334.45	173	2450	9204	6290	6776.24
1991	129	475.05	402.45	207	2760	10934	7550	7972.04
1992	144	516.85	432.45	233	2850	11804	7795	8469.04
1993	172	722.7	626.53	354	3720	16556	10464	11638.8
1994	182	760.25	662.58	382	3970	17586	10834	12331.8
1995	192	885.05	783.36	451	4690	20506	12429	14391
1996	198	924.75	815.86	484	4951	21721	12829	15080.94
1997	215	999.8	883.41	534	5127	25916	14029	17415.28
1998	239	1096.3	955.96	601	5467	28311	14864	18868.18
1999	253	1163.1	1021.16	673	5682	30263	15499	20005.3
2000	269	1235.15	1084.61	724	5947	31433	16129	20813.9
2001	276	1271.45	1119.91	751	5947	32438	16644	21358.6
2002	278	1293.2	1136.66	769	5997	33073	16769	21702.7
2003	283*	1305	1148.46	779	5997	33548	16964	21952.4

*Information for one farmer who started his fertiliser use on groundnut and diffused it to jowar, cotton, maize is not available because of recall failure.

Three other facts need to be appreciated within the following constraints faced by the sample farmers: First, it's only after 1989, urea use significantly increases (Table 5.7 and Chart 5.1), otherwise, between 1970 and 1989, there is no significant difference in diffusion of urea, DAP and SSP. Second, from an average 1.5 crops fertilised till 1989, the average number of crops fertilised by farmers has increased to three crops (2.8). Lastly, in the total nutrient use, as the diffusion process reveals, there is high nutrient imbalance with no use of potash by any of the sample farmers. The phosphate and nitrogen ratio even in 2003 is 1:3 rather than 1:2. Over the period of time, there is significant variation in diffusion process across crops (Appendix 5.2). From 163 first-time adopters on maize, fertiliser use on the crop has grown to 217 farmers. Over the period of time fertiliser use on maize has diffused to 72 per cent of the sample cultivators. While diffusion on maize (72%), cotton (54%), jowar (41%) and wheat (34%) is significant, in the crops like kharif pulses, gram and bajra, less than 5 per cent farmers have diffused fertiliser use. The diffusion of fertiliser use on soyabean, groundnut and chillies is also low.

Variations in fertiliser use by first-time users reveals that it is basically the diffusion of use, in the entire cropped area that accounts for total variations of first-time user. About half of fertiliser nutrients used by first time fertiliser users were applied on maize, followed by cotton (26%), jowar (14%) and wheat (8%). As been already stated there were marginal variations in the rates of fertiliser use (Table 5.2) but total area fertilised by maize growers is about three times higher than jowar area fertilised by the growers and about six times higher than cotton and wheat area fertilised. Thus, the variations across crop share, whereby maize occupying the highest place is accounted by large number of farmers applying fertilisers in the crop in relatively large area through at relatively low rates.

As Table 5.1 has revealed that fertiliser adoption on a crop has been with considerable time lag. Even in 2003 we observed there were five farmers who started fertiliser use first time. On the other hand, even in 2003, 18 (about 6%) farmers are non-adopters of fertilisers. In the subsequent section, an attempt is made to understand the variations in fertiliser adoption among the sample farmers.

Time Lag in Fertiliser Adoption

A number of factors, including farm characteristics, are identified by researchers that contribute to variations in time of adoption of fertiliser use (Desai and Singh, 1973). Reconstruction of these farm characteristics for past is difficult if not impossible. What is, however, easily and reliably available in terms of farm characteristic, is the current status of farmers. As bulk of fertiliser adoption has taken place in 1990s, it will not be highly unrealistic to explain time lag in fertiliser adoption using the contemporary characteristics of the farms.

Table 5.8 presents the distribution of sample farmers across time lag of fertiliser adoption and the income profile of

Table 5.8
Distribution of Farmers across Income and Adoption Year Categories

Fertiliser use adoption category	*Income rupees per farm*			*All*
	Up to 9000	*9001 to 17000*	*> 17000*	
	Number of farmers			
Non-users	12	5	1	18
Adopted 5 years back	29	22	17	68
Adopted 6 to 15 years back	46	48	38	132
Adopted more than 15 years	12	27	44	83
All	99	102	100	301

Chi-square = 33.57, Significance 0.000.
r = +0.32, Significance 0.000.

Source: Based on the data collected by the author.

the farmers. A scrutiny of the table reveals that those farmers whose income is relatively high are also farmers who have adopted fertiliser use relatively early whereas those farmers whose income is low have delayed adoption of fertiliser use. The visual scrutiny of Table 5.8 is substantial by the chi-square test and the correlation between the two variables (Table 5.9). A high and significant chi-square and a positive correlation reveal that there is strong positive association between income of farmers and their fertiliser adoption behaviour. Those with relatively higher income have had adopted fertiliser use relatively early.

Table 5.9

Association of Fertiliser Adoption Behaviour with Some Explanatory Factors

Explanatory factor with category	*Chi-square*	*Significance*	*Correlation*	*Significance*
Education (nil to high)	8.1	0.046	+0.12	0.020
Landholding (small to large)	14.3	0.026	+0.22	0.000
Irrigation (low to high)	21.6	0.001	+0.23	0.000
Holdings spread in numbers (few to large)	13.8	0.032	0	NS
Migration (no and yes)	10.5	0.015	–0.18	0.002
Size of Family (small to high)	1	NS	0	NS
Income (low to high)	33.6	0.000	+0.32	0.000
Dependency on Agriculture (low to larger % of income coming from agriculture)	53.9	000	+0.33	0.000
Soil type (thin rocky to deep soil)	36	000	+0.29	0.000
Level surface (small to large proportion)	31	0.000	+0.27	0.000
Tribes (ST, SC, Patidar)	1	NS	0	NS
Tenancy (no and yes)	1	NS	0	NS

Similar analysis has been attempted where adoption process is associated with (a) level of education, (b) size of landholding, (c) soil type, (d) proportion of plain area available for cultivation, (e) tenancy status, (f) level of irrigation, (g) family size, (h) dependency on agricultural income, and (i) tribe (Table 5.8).

Let us begin with tribe, tenancy, size of the family and number of holdings. What we observe is that none of these variables influences the fertiliser adoption behaviour of sample farmers. Irrespective of the fact that whether one is a tenant or landlord, one has lease land in or leased land out, his fertiliser adoption behaviour is same. On the other hand, soil type, condition of land, migration pattern, level of education, size of holding, proportion of GCA irrigated, total income and proportion of agricultural income in total significantly influence the observed fertiliser adoption behaviour.

What the above findings imply? These findings indicate that none of the social processes including tenancy is influencing the fertiliser adoption behaviour. On the other hand, economic indicators like better soil condition, higher proportion of plain agricultural land, non-migrant, relatively educated head of the household, large farmers, farmers with more irrigation, rich farmers and farmers whose proportion of agricultural income in total income is higher are relatively early fertiliser adopters. On the other hand, relatively large number of farmers with adverse conditions like low income, who migrate, have poor soil conditions, are uneducated, have low income and high dependency on migration and labouring are non-adopters or late adopters of fertiliser use.

These findings are relevant in many ways. First, they indicate that farm condition rather than farmer's condition influences fertiliser adoption behaviour. Second, those engaged in livelihood struggles that distract them from their agriculture like relatively more labouring around the village and migration, are poor agriculturists on whose farm technology transfer in the form of fertiliser use is constrained by lack of resources. Third, better farm conditions do positively influence adoption of fertiliser use and hence avenues for agricultural technology transfer. Lastly, diffusion of fertiliser use to other crops after its adoption is also significantly influenced by the economic processes rather social conditions.

Summing up

Unlike intensively irrigated areas where adoption of fertiliser use has long history, in our sample of tribal farmers, adoption of fertiliser use started as late as 1970. The most important crop chosen for fertiliser use by early adopters was maize. Till 1980, only a few farmers adopted fertiliser use. It was only between 1992 and 2003 that there was sizable increase in fertiliser adoption; it took over 20 years for first 50 per cent of farmers to adopt fertilisers, while in next 10 years rest 50 per cent farmers adopted fertilisers. The early adopters started fertiliser use with one of the following crops: maize, cotton, wheat, groundnut and chilli. The late adopters started fertilising their crops with either soyabean or bajra. Interestingly, bulk of early fertiliser adopters on maize were unirrigated farmers. A sizable area under irrigation in crops that was fertilised was in the case of soyabean, wheat and cotton. For rest of the crops, less than 8 per cent of the crop area under fertilisers was irrigated. When fertiliser use was adopted first time on groundnut, chillies and bajra, it was adopted on totally unirrigated area. Urea was the main fertiliser used by the farmers when fertilisers were first adopted, followed by Single Super Phosphate and Diammonium Phosphate. None of the farmers used potash when applied fertilisers first time. In terms of nutrient use per unit of area, highest use by first-time adopters was on cotton, 34 kg per fertilised acre. The intensity of fertiliser use on the crop maize, on which maximum number of farmers adopted fertiliser was only 14.6 kg per acre. Surprisingly, fertiliser use on two other irrigated crops soyabean and wheat by first-time adopters was only 16 kg per acre and 13.7 kg per acre respectively. In fact, except for cotton, there is no significant variation in fertiliser use per unit of area across crops by first-time users. This indicates that the farmers used fertilisers cautiously when they fertilised their crops first time; use was confined only to one crop and at reasonably low rates.

The tribal farmers were well aware of the nutrient deficiency but the physical condition of their land turned them into late adopters of fertilisers. The experiences of first-time adopters of fertiliser use have been quite encouraging but not without dissent; increased production due to fertiliser use, improvement in quality of production and crop maturing early are the most frequently reported experiences given by a substantial number of the farmers. A minuscule of farmers did report negative experiences like poor crop response, rapid loss of moisture owing to fertiliser use and harmful effects on crop. But, positive effects of fertiliser use dominated the tribal psyche.

After adopting fertiliser use on maize, 163 farmers diffused it to 299 other crops giving average less than two crops fertilised by them over time. In the case of five growers who adopted fertiliser for the first time on groundnut, diffused fertiliser use to another 12 crops giving an average additional 2.5 crops fertilised over time. This means that for some crops like maize a large number of farmers adopted fertilisers with small time lag but diffused fertiliser use to other crops slowly. For groundnut and wheat, though adoption of fertiliser was relatively late, the diffusion of fertiliser use to other crops was relatively faster. This contrast in diffusion process emerges clearly when we observe that less than 16 per cent of growers who adopted fertilisers on maize, cotton or jowar, did not diffuse fertilisers to pulse and bajra at all. Evidently, a very large proportion of growers diffused fertiliser use on maize, cotton and wheat when they adopted fertilisers with bajra.

Observing the adoption and diffusion of fertiliser use on three important crops maize, jowar and cotton, it is noteworthy that while early adopters fertilised mainly maize, jowar, cotton and wheat, and added over the years soyabean, bajra, gram and sometimes kharif pulses, the late adopters mainly started fertiliser use with maize, cotton, soyabean, chillies, wheat and kharif pulses. It is only after 1989 that urea use significantly increased, otherwise between 1970 and 1989, there was no significant difference in diffusion of urea, DAP and SSP. From an average 1.5 crop fertilised till 1989, the average number of crops fertilised by farmers increased to 3 crops. Nevertheless, there is high nutrient imbalance with no use of potash by any of the sample farmers. The phosphate and nitrogen ratio even in 2003 is 1:3 rather than 1:2. Over the period of time, there is significant variation in diffusion process across crops. While diffusion of fertiliser use on maize (72%), cotton (54%), jowar (41%) and wheat (34%) is significant in the crops like kharif pulses, gram and bajra, less than 5 per cent farmers have diffused fertiliser use. The diffusion of fertiliser use on soyabean, groundnut and chillies is also low. If we assume that first-time adopters kept on fertilising their the crops at the same level over time, the diffusion of fertiliser use resulted in a total NPK use of about 22,000 kg by the sample farmers.

None of the social processes including tenancy is influencing the fertiliser adoption behaviour. The economic indicators like better soil condition, higher proportion of plain agricultural land, non-migrant households, relatively educated household, large land holdings, farmers with more irrigation, rich farmers and farmers whose proportion of agricultural income to total income is higher are relatively early fertiliser adopters. These findings are relevant in many ways. First, they indicate that farm condition rather than farmer's condition influences fertiliser adoption behaviour. Second, those engaged in livelihood struggles that distract them from their agriculture like relatively more labouring around the village and migration, are relatively poor agriculturists on whose farm technology transfer in the form of adoption of fertiliser use is relatively constrained due to lack of resources. Third, better farm conditions do positively influence adoption of fertiliser use and hence avenues for agricultural technology transfer. Lastly, diffusion of fertiliser use to other crops after its adoption is also significantly influenced by the economic processes rather social conditions.

APPENDIX 5.1
Diffusion of Fertiliser in Different Crops after Adoption Over time

Fertiliser beginning with maize	*Fertilised crops between different years*									
	Before 1985		*1985–90*		*1991–95*		*1996–2000*		*2001–2003*	
Before 1985 (24)	Jowar	2	Jowar	5	Cotton	12	Soyabean	1	—	—
	Cotton	2	Wheat	5	Jowar	4	Cotton	1		
	Groundnut	1	Cotton	2	Chilli	4	K. Pulses	2		
	Wheat	1			Soyabean	3	Groundnut	1		
					Wheat	2	Chilli	2		
					Groundnut	2				
					Bajra	1				
					K. Pulses	1				
Total		6		12		29		7		—
1985–90 (61)	—	—	Wheat	8	Cotton	23	Cotton	15	Cotton	1
			Cotton	6	Wheat	15	Wheat	5	Groundnut	1
			Chilli	4	Jowar	9	Jowar	4	Bajra	1
			Jowar	3	Soyabean	7	Soyabean	4		
					Chilli	4	Groundnut	4		
					Groundnut	4	Chilli	2		
					Gram	2	Gram	2		
					Bajra	2	Bajra	1		
					K. Pulses	3	K. Pulses	2		

Cont'd...

Cont'd...

Total		—		21		69		39		3
1991–95 (30)	—	—	—	—	Cotton	12	Cotton	7	Cotton	1
					Wheat	9	Jowar	4	Wheat	1
					Jowar	8	Soyabean	3		
					Bajra	3	Wheat	1		
					Soyabean	2	Chilli	1		
					Chilli	2	Bajra	1		
					Groundnut	1				
Total		—		—		37		17		2
1996–2000(42)	—	—	—	—	—	—	Cotton	13	Cotton	4
							Soyabean	8	Jowar	3
							Jowar	7	Soyabean	2
							Wheat	7	Gram	1
							K Pulses	2	Chilli	1
							Chilli	1		
Total		—		—		—		38		11
2001–2003 (6)	—	—	—	—	—	—	—	—	Chilli	3
									Wheat	2
									Jowar	2
									Groundnut	1
Total		—		—		—		—		8

Cont'd...

Cont'd...

Fertiliser beginning with jowar	*Fertilised crops between different years*									
	Before 1985		*1985–90*		*1991–95*		*1996–2000*		*2001–2003*	
Before 1985 (0)	—	—	—	—	—	—	—	—	—	—
Total		—		—		—		—		—
1985–90 (14)	—	—	Maize	2	Cotton	5	Cotton	3	Cotton	1
					Maize	5	Soyabean	1		
					Wheat	4	Groundnut	1		
					Bajra	2	Chilli	1		
					Groundnut	2	Wheat	1		
					Gram	2	Bajra	1		
					Soyabean	1				
					K. Pulses	1				
Total		—		2		20		8		1
1991–95(26)	—	—	—	—	Maize	5	Bajra	6	—	—
					Cotton	4	Maize	3		
					Wheat	4	Cotton	3		
					Groundnut	3	Wheat	2		
					Bajra	1	Soyabean	1		
					Soyabean	1				

Cont'd...

Cont'd...

Total		—		—		18		15		—
1996–2000 (15)	—	—	—	—	—	—	Cotton Bajra Groundnut	4 4 3	Bajra Maize	2 1
Total		—		—		—		11		3
2001–2003 (3)	—	—	—	—	—	—	—	—	Maize Bajra Groundnut	1 1 1
Total		—		—		—		—		3
Fertiliser beginning with cotton	*Fertilised crops between different years*									
	Before 1985		*1985–90*		*1991–95*		*1996–2000*		*2001–2003*	
Before 1985 (6)	Maize	2	Maize Jowar Chilli Wheat Soyabean	3 2 2 1 1	Chilli Wheat Bajra K. Pulses	2 2 1 1	—	—	—	—
Total		2		9		6		—		—

Cont'd...

Cont'd...

1985–90 (5)	—	—	Maize	2	Wheat	2	Maize	2	—	—
					Jowar	1	Bajra	1		
							Wheat	1		
							Gram	1		
Total		—		2		3		5		—
1991–95 (6)	—	—	—	—	Jowar	3	Maize	1	Wheat	2
					Maize	3	Wheat	1		
					Groundnut	1	Jowar	1		
							Groundnut	1		
							K. Pulses	1		
Total		—		—		7		5		2
1996–2000 (4)	—	—	—	—	—	—	Maize	3	Wheat	2
							Jowar	2	Jowar	1
							Bajra	1		
							Groundnut	1		
Total		—		—		—		7		3
2001–2003 (4)	—	—	—	—	—	—	—	—	Maize	1
									Chilli	1
Total		—		—		—		—		2

Cont'd...

Cont'd...

Fertiliser beginning with wheat	*Fertilised crops between different years*									
	Before 1985		*1985–90*		*1991–95*		*1996–2000*		*2001–2003*	
Before 1985 (3)	—	—	Maize	2	Maize	1	Groundnut	1	Cotton	1
			Cotton	1	Cotton	1	Cotton	1		
					Jowar	1				
					Soyabean	1				
Total		—		3		4		2		1
1985–90(5)	—	—	Cotton	1	Maize	2	Cotton	1	Chilli	1
					Bajra	1				
					Jowar	1				
Total		—		1		4		1		1
1991–95 (4)	—	—	—	—	Cotton	1	Jowar	3	Maize	1
							Maize	2		
							Cotton	1		
							Chilli	1		
Total		—		—		1		7		1

Cont'd...

Cont'd...

1996–2000 (7)	—	—	—	—	—	—	Maize	5	Bajra	1
							Cotton	4	Soyabean	1
							Jowar	3		
							Bajra	2		
							Chilli	2		
							K. Pulses	2		
							Groundnut	1		
Total		—		—		—		19		2
2001–2003(0)	—	—	—	—	—	—	—	—	—	—
Total		—		—		—		—		—
Fertiliser beginning with Bajra	*Fertilised crops between different years*									
	Before 1985		*1985–90*		*1991–95*		*1996–2000*		*2001–2003*	
Before 1985 (0)	—	—	—	—	—	—	—	—	—	—
Total		—		—		—		—		—
1985–90 (1)	—	—	—	—	—	—	—	—	—	—
Total		—		—		—		—		—
1991–95 (2)	—	—	—	—	—	—	Wheat	2	Chilli	1
							Maize	1		
Total		—		—		—		3		1

Cont'd...

Cont'd...

1996–2000 (6)	—	—	—	—	—	—	Maize Gram Wheat Cotton Jowar	2 1 1 1 1	Jowar	1
Total		—		—		—		6		1
2001–2003 (1)	—	—	—	—	—	—	—	—	—	—
Total		—		—		—		—		—

Fertiliser beginning with groundnut	*Fertilised crops between different years*									
	Before 1985		*1985–90*		*1991–95*		*1996–2000*		*2001–2003*	
Before 1985(0)	—	—	—	—	—	—	—	—	—	—
Total		—		—		—		—		—
1985–90 (2)	—	—	Maize Soyabean Wheat	1 1 1	—	—	—	—	—	—
Total		—		3		—		—		—

Cont'd...

1991–95 (2)	—	—	—	—	Chilli	1	Wheat	2	—	—
					Maize	1	Soyabean	1		
					Cotton	1	Chilli	1		
							Cotton	1		
Total		—		—		3		5		—
1996–2000 (1)	—	—	—	—	—	—	Cotton	1	—	—
Total		—		—		—		1		—
2001–2003 (0)	—	—	—	—	—	—	—	—	—	—
Total		—		—		—		—		—
Fertiliser beginning with soyabean	*Fertilised crops between different years*									
	Before 1985		*1985–90*		*1991–95*		*1996–2000*		*2001–2003*	
1996–2000 (2)	—	—	—	—	—	—	Maize	1	—	—
							Wheat	1		
							Cotton	1		
2 Total		—		—		—		3		—
Fertiliser beginning with chilli	*Fertilised crops between different years*									
	Before 1985		*1985–90*		*1991–95*		*1996–2000*		*2001–2003*	

K. Pulses: Kharif Pulses

APPENDIX 5.2
Number of Farmers Diffusing Fertiliser Use Over Time

Year	*Maize*	*Soyabean*	*Jowar*	*Cotton*	*K. Pulses*	*Groundnut*	*Wheat*	*Gram*	*Chilli*	*Bajra*
					Number of Farmers					
2003	—	—	1	2	—	1	2	—	3	1
2002	4	1	1	4	—	2	1	—	2	3
2001	7	2	6	6	—	—	1	1	2	2
2000	9	4	11	9	1	4	6	1	—	6
1999	18	5	9	21	3	4	5	—	3	4
1998	21	6	8	12	2	1	5	1	4	7
1997	6	2	9	14	2	3	6	—	4	4
1996	6	3	6	5	1	2	10	—	—	—
1995	11	4	10	13	3	2	15	1	5	5
1994	4	4	4	9	—	—	5	—	1	1
1993	22	4	25	28	1	10	18	1	6	6
1992	8	—	5	4	—	3	5	—	—	1
1991	6	4	5	11	1	—	6	—	1	—
1990	24	—	10	7	—	—	1	1	1	1
1989	5	—	1	—	—	—	P1	—	—	—

Cont'd...

Cont'd...

1988	23	1	9	8	—	—	9	1	5	—
1987	5	—	1	1	—	—	—	—	—	—
1986	4	—	—	2	—	—	—	—	—	—
1985	8	—	—	—	—	1	4	—	—	—
1984	2	—	—	—	—	—	—	—	—	—
1983	12	—	2	4	—	—	2	—	—	—
1982	1	—	—	—	—	—	—	—	1	—
1981	2	—	—	—	—	—	1	—	—	—
1980	6	—	—	1	—	—	—	—	—	—
1978	2	—	—	1	—	—	—	—	—	—
1970	1	—		—	—	—	—	—	—	—
Total	217	40	123	162	14	33	102	7	28	4

K. Pulses: Kharif Pulses

Chapter 6

Growth in Fertiliser Use between 2001–02 and 2003–04

The objectives of this chapter are three-fold. First, to estimate how the patterns of NPK use amongst 283 sample farmers have changed between 2001–02 and 2003–04. Second, to examine the sources of growth in NPK use on ten major crops that constitute the cropping pattern of sample farmers. Thirdly, to estimate and examine the untapped potential of fertiliser use. The conceptual frame that helps us analyse these changes has been elaborated in Chapter 3. To recapitulate, the growth in NPK use between 2001–02 and 2003–04 can be decomposed into growth in NPK use because of continuous users who have used fertilisers all the three years + growth because of continuous years of 2001–02 and 2002–03 – dropout of 2002–03 + Adopters of year 2002–03 + growth because of continuous users of 2002–03 and 2003–04 – dropouts of year 2003–04 + Adopters of year 2003–04. The analysis in this chapter has been divided in five sections. The first section presents different patterns of fertiliser use. The second section presents the estimates of growth in fertiliser use. The third section examines the sources of growth in fertiliser use. The fourth section tries to estimate and examine untapped potential of fertiliser use. The last section presents a summary of the major findings.

I

Pattern of Fertiliser Use

Table 6.1 gives the number of farmers falling under different patterns of fertiliser use for all the crops taken together as well as for different crops between 2001–02 and 2003–04. Although, fertiliser use is taking place on 283 farms, seven farmers, though had adopted fertiliser use earlier, have not applied fertilisers between 2001–02 and 2003–04. Another 18 farmers are non-adopters of fertilisers even in 2003–04. Interestingly, seven farmers applied fertilisers only one out of the three years, fourteen farmers applied fertilisers only two out of the three years and 255 farmers used fertiliser all the three years. Thus, though at the farm level only seven cultivators have not used fertilisers during all the three years, the crop-specific non-users in all the three years vary significantly. The non-users are relatively low in maize (29%), cotton (46%), jowar (57%) and wheat (66%). For the remaining crops, proportion of non-users ranges from 84 per cent to 99 per cent; this is very high rate of non-adoption by any standard. The other important point emerging from the Table 6.1 is the nature of discontinuation of fertiliser use amongst the farmers. For all the crops taken together, though the proportion of infrequent users is just 8 per cent of the number of farmers that use fertilisers continuously all the three years, for some individual crops this proportion is significantly large; this is especially true in the case of gram, kharif pulses, groundnut, bajra and soyabean. It means that for these crops not only the total number of farmers adopting fertiliser is

Table 6.1

Number of Farmers using Fertilisers between 2001–02 and 2003–04

	Non-user all three years	*Only first year users*	*Used only last year*	*Used only middle year*	*Used last two years*	*Used only first and second year*	*Used only first and last year*	*Used all the three years*			
								Fluctuating	*Decreasing*	*Same*	*Increasing*
All the crops	7	—	5	2	7	4	3	132	3	106	14
Maize	81	5	5	1	7	10	4	38	—	129	3
Jowar	162	7	2	7	10	10	7	12	—	64	2
Cotton	131	9	8	4	6	6	3	43	1	72	—
Soyabean	237	6	2	1	5	1	2	14	—	15	—
Groundnut	246	4	3	10	2	5	—	3	—	13	—
Wheat	188	13	6	—	4	—	3	18	—	48	3
Gram	281	—	1	—	—	—	—	—	—	1	—
Chillies	251	5	—	—	4			6	—	15	2
Bajra	226	8	13	2	3	1	2	6	—	21	1
Kharif pulses	273	1	3	3			1	1	—	1	—

Source: Tables in the chapter are based on primary data collected by the author.

relatively low but also there is a high rate of dropout of fertiliser use between 2001–02 and 2003–04.

The implications of this pattern shall be seen in growth of fertiliser use as well as its sources. If non-users for all the three years are high then neither the growth shall be high nor any one of the sources could be dominating in the growth process. But if non-users during all the three years are relatively few, not only growth shall be substantial, it is also likely that continuous users rather than new adopters could dominate the growth. Table 6.2 gives NPK use on different crops between 2001–02 and 2003–04. Fertiliser use has grown from 24,800 kg in 2001–02 to 28,400 kg in 2002–03. The use thereafter stagnated at 28,100 kg in 2003–04. The sluggishness in use in 2003–04 has been an outcome of fall in use in all the crops, except chillies and soyabean. Despite the sluggish growth and fall in NPK use, three crops—cotton, wheat and jowar—account for two-thirds of total use of NPK. In this, cotton alone accounts for about a third of total use. Despite stagnation and fall in NPK use during 2003–04, the use in the year is still higher in comparison to 2001–02. The increase in NPK use, between 2001–02 and 2003–04, on three crops—cotton, chillies and wheat—accounts for about two-thirds of growth in fertiliser use. Kharif pulses and groundnut registered a decline in fertiliser consumption while growth in fertiliser use on gram was nil.

There seems to be a significantly large untapped potential for fertiliser use owing to non-use on three major growth generating crops; more than half of cultivators do not fertilise cotton, over two-thirds of farmers do not apply fertilisers on wheat and about nine out of ten cultivators do not apply fertiliser on chillies. This is partially true, for at this stage, we do not know how many of 131 non-users on cotton have had grown cotton at all? This shall be analysed in next chapter. We, at the moment, are interested in estimating and examining relative importance of different sources of growth in fertiliser

Table 6.2
Crop-specific NPK Consumption

Crop	*2001–02*	*2002–03*	*2003–04*
		Kilogram	
Maize	4,812.36	5,318.22	5,048.84
Jowar	2,439.96	2,689.64	2,630.68
Cotton	7,972.30	9,461.50	8,931.00
Soyabean	2,161.22	1,549.16	2,310.58
Pulses	56.10	101.10	41.40
Groundnut	793.70	750.60	653.06
Wheat	3,541.70	4,203.10	4,414.70
Gram	216.00	436.00	216.00
Chillies	2,405.98	3,259.80	3,360.96
Bajra	423.46	608.61	519.86
All crops	24,822.78	28,377.73	28,127.08

use on different crops. What is actually the growth in fertiliser use and what are its sources for different crops shall be seen in the next section.

II

Growth in Fertiliser Use

Table 6.3 presents average annual fertiliser use for ten crops on which fertilisers have been used by sample farmers between 2001–02 and 2003–04. Following important points are worth noting. About 87 per cent of average NPK use is accounted by the farmers who continuously used fertilisers all the three years. This proportion is still larger in the case of gram, cotton and maize. On the other hand, pulse, bajra, jowar and groundnut had shown a tendency whereby growth generating contribution of continuous users has been significantly low and the contribution to growth-constraining users who had

Table 6.3

Average Per Annum NPK Use in Kilogram under Different Patterns

Crop	*Non-user all three years*	*Only first year users*	*Used only last year*	*Used only middle year*	*Used last two years*	*Used only first and second year*	*Continuous user all the three years*	*Total*
Maize	44.43	31.43	3.83	184.73	228.17	26.83	4540.37	5059.81
Jowar	69.50	19.17	73.30	238.20	155.37	85.41	1945.82	2586.76
Cotton	75.17	65.50	42.20	183.23	71.37	79.67	8271.13	8788.27
Soyabean	52.82	14.50	10.67	67.80	10.33	51.83	1799.03	2007.
K. Pulses	3.07	22.23	12.97	–	—	1.53	24.00	66.20
Groundnut	9.32	30.37	92.07	82.87	16.00	—	501.83	732.45
Wheat	176.33	138.80	—	102.67	—	110.47	3524.90	4053.17
Gram	—	73.33	—	—	—	—	216.00	289.33
Chillies	20.30	—	—	114.17	—	—	2874.45	3008.91
Bajra	47.27	67.42	15.33	21.47	15.13	18.40	332.29	517.31
All crops	498.21	462.75	250.37	995.14	496.37	374.14	24029.82	27109.2

dropped out of fertiliser use is significantly high. Those farmers who had discontinued fertiliser use—like users of only 2001–02 or users of only 2002–03 or users of 2001–02 and 2002–03—create a retarding affect on growth of fertiliser use. Although these growth-retarding farmers have had only 4 per cent contribution to total use, on individual crops like groundnut, jowar, bajra and maize this proportion has sizeable dampening effect on growth of fertiliser use. The growth of fertiliser use, however, shall also be affected by fluctuating use of continuous users in a crop. What the above findings indicate is that though continuous users may dominate the use, the growth shall be significantly affected by the both the patterns of use of continuous users as well as the discontinuous users.

Non-adoption of fertiliser use and discontinuous use of fertilisers, if takes place owing to not applying fertilisers rather than not growing the crop, also indicates that a significant proportion of potential of fertiliser use on the crop remains untapped. Reading the Table 6.1 along with Table 6.3 it can be concluded that a high proportion of potential remains untapped in the case of gram, pulses, chillies, groundnut, bajra and soyabean owing to non-adoption of fertilisers. On the other hand, number of farmers discontinuing fertiliser use significantly affects growth in fertiliser use on all the crops except gram and chillies. In the case of groundnut, 15 per cent of average use is by farmers that discontinued fertiliser use. The contribution of growth retarding discontinuation of fertiliser use is 9 per cent in the case of jowar, 6 per cent in the case of bajra and 5 per cent in the case of maize. For all these three crops, a sizable proportion of deceleration in fertiliser use shall be owing to infrequent use.

This is not surprising, for in the initial years the growth in effective demand remains low because only the most enterprising farmers and those with the best production functions adopt fertiliser use (Desai, 1969). Even under conditions of slow growth of fertiliser use, some farmers start using fertilisers on fertiliser-responsive crops like wheat, cotton and chillies. The rate of growth in fertiliser use, therefore, shall be low initially and then accelerate gradually. The best of the experiences of some successful entrepreneurs help in demonstrating to others that risk of fertiliser use is low in these crops than others. If farmers face risk in fertiliser application on some difficult crops that have low potential, they are likely to discontinue fertiliser use for some time. Depending on perceptions, this process may vary across crops making the sources of growth in fertiliser use different for different crops. It will be worth understanding which of the sources of growth in fertiliser use at the micro level dominate in different crops. The quantitative dimensions of various sources of growth in fertiliser use are discussed below.

III

Different Sources of Growth

The NPK use of sample farmers between the three points of time can be categorised into seven different patterns as conceptualised in the beginning of the chapter. We can further categorise the ten crops that we are analysing for their sources of growth, in three major categories. Patterns in the growth-generating crops wheat, cotton and chillies; patterns in dominant crops maize and jowar; and pattern in low growth crops bajra, soyabean, kharif pulses, gram and groundnut. The growth in NPK use in different crops and the sources of this growth are presented in Table 6.4. The relative importance of different sources of growth in NPK use is given in Table 6.5.

Findings of these tables can be summarised as follows: First, growth pattern reveals that in crops like wheat, chillies and cotton, total increase in NPK use was substantial; in crops like maize, jowar, soyabean and bajra the growth in NPK use was moderate while in the case of pulse, groundnut and gram

Table 6.4
Decomposition of Growth in NPK Use into its Components

Crop	*Growth because of continuous users*	*Growth due to continuous users of year 1 & 2*	*Dropouts of year 2*	*Adopters of year 2*	*Continuous users of year 2 &3*	*Dropouts of year 3*	*Adopters of year 3*	*Total growth in NPK use between 2001–02 and 2003–04*
Maize	124.38	46.5	–50.6	341.5	59.8	–459.5	174.7	236.48
Jowar	93.5	–25.5	–337.2	439.3	–49	277.8	347.42	190.72
Cotton	657.3	–21.1	–2461.1	426.1	90.5	–293	345	958.7
Soyabean	–53	–15	–112	143.5	3.4	–51.5	233.96	149.36
Pulses	15	—	–41.2	66.7	—	–66.7	11.5	–14.7
Groundnut	9	0	–276.2	217.1	–3.4	–115.1	27.96	–140.64
Wheat	143	—	–150.2	554.4	32	–416.4	710.2	873
Gram	0	—	—	220	—	–220	—	0
Chilli	765.58	—	—	214	–85.5	—	60.9	954.98
Bajra	–25.3	–8.6	–66.7	227.56	13.8	–220.66	176.3	96.4
All crops	3030.5	0	–69	260.9	8.1	–147.2	221	3304.3

Table 6.5
Relative Importance of Different Sources of Growth in NPK Use

Crop	Growth because of continuous users	Growth due to continuous users of year 1 & 2	Dropouts of year 2	Adopters of year 2	Continuous users of year 2 &3	Dropouts of year 3	Adopters of year 3	Total growth in NPK use between 2001–02 and 2003–04
High Growth-Generating Crops								
Wheat	16.4	—	–17.2	63.5	3.7	–47.7	81.3	100 (873)
Cotton	68.6	–2.2	–25.7	44.4	9.4	–30.6	36.0	100 (959)
Chilli	80.2	—	—	22.4	–8.9	—	6.3	100 (955)
Moderate Growth-Generating Crops								
Maize	52.6	19.7	–21.4	144.3	25.3	–194.4	73.9	100 (236)
Jowar	49.0	–13.4	–176.8	230.3	–25.7	–145.7	182.3	100 (191)
Soyabean	–35.5	–10.0	–75.0	96.0	2.3	–34.5	156.7	100 (149)
Bajra	–26.2	–8.9	–69.2	236.0	14.3	–228.9	182.9	100 (96)
Deceleration-Generating Crops								
Pulses	–102.0	—	280.3	–453.7	—	453.6	–78.2	100 (15)
Groundnut	–6.4	0	196.3	–154.4	2.4	81.0	–19.9	100 (141)
Gram	0	—	—	+ a	0	–a	0	100 (0)
All crops	91.7	0	–2.1	7.9	0.2	–4.4	6.7	100 (3304)

a: Insignificant growth
Figures in parentheses are growth in NPK use in between 2001-02 and 2003-04.

the growth in NPK use was negative or stagnant. Second, users continuously applying fertilisers all the three years contribute to total growth in NPK (a) substantially in the case of chillies and pulses, (b) moderately in the case of maize and jowar, (c) low in the case of groundnut, and (d) negative in the case of bajra and soyabean. Third, a substantial reduction in growth in fertiliser use has taken place owing to discontinuation of use in the case of maize and bajra, during 2003–04 and in the case of jowar, pulse and groundnut for both 2002–03 and 2003–04. Fourth, a substantial growth in NPK use during 2003–04 has been achieved in the case of jowar, soyabean, wheat and bajra. Lastly, the major source of growth varies significantly across groups of crops: (a) For wheat it is new adopters that created substantial growth in NPK use; in the case of cotton and chillies it is continuous users as well as new adopters that generated growth in NPK use; (b) in the case of maize, jowar, soyabean and bajra, the new adopters did substantially contribute to growth in NPK use; and (c) in the case of low growth crops kharif pulses, groundnut and gram, low contribution of continuous user as well as cancelling out of effect of dropout and new adopters resulted in low growth in NPK use. The last observation needs more explanation, for through this understanding one can articulate why fertiliser use does not grow rapidly in low fertiliser using areas. Analysing growth in NPK use this way may also provide us valuable insights on how to increase fertiliser use in the remote tribal areas.

Chillies among the three growth generating crops is the only case that has demonstrated that growth in NPK use took place due to continuous use of NPK between 2001–02 and 2003–04. The other sources of growth, in the case of chillies, played only marginal role. On the other hand, untapped potential, despite a sizeable growth in NPK use between 2001–02 and 2003–04 on wheat and cotton, has been owing to discontinuation of use in 2002–03 as well as 2003–04. About 26 per cent additional growth in use could have been easily achieved in case of cotton if the discontinuation of use could have been avoided in 2002–03. In the case of wheat, about 48 per cent additional growth in NPK use could have been achieved, had there been less discontinuation of fertiliser use in 2003–04. The moderate growth group crops, maize and jowar, which have relatively high adoption of fertiliser, fail to register high growth owing to significantly large discontinuation of fertiliser use in 2002–03 (as the case is with respect to jowar) or during 2003–04 (in the case of maize). For the other crops in this group, bajra and soyabean, the growth in continuous users was negative. It can be safely concluded that for maize, jowar, soyabean and bajra, the growth in NPK use could be achieved owing to new adopters especially during 2003–04. However, in the case of these moderate growth crops, untapped potential is significantly high owing not only to discontinuous use but also owing to relatively moderate use by continuous users. For negative growth crops—pulses, groundnut and gram—not only growth in the case of continuous users of fertilisers is low, but these crops also have high discontinuation of fertiliser use. In this group of crops, not only rate of adoption of fertiliser is low and the continuous use is contributing marginally but also the discontinuous use is sometimes higher even compared to new adopters. This results in negative or low growth in NPK use. In a sense, the use pattern on kharif pulses, gram and groundnut at different levels shows a high untapped potential.

But before conclusively accepting the nature of untapped potential the way we have developed, it will be necessary to ascertain if unfertilised crops as reported above were in fact grown by the farmers? Equally important is to understand why non-adoption, during all the three years, is so high in the crops like bajra, soyabean, kharif pulses, gram and groundnut. Understanding the reasons of discontinuation of fertiliser use by maize and jowar growers may also be an important input to increasing fertiliser use. On the other hand, it is important to know why the number of farmers who are continuously using fertilisers in maize and jowar is low. The analysis in the next section tries to answer these questions.

IV

Untapped Potential for Fertiliser Use

The objective of this section is first to estimate and then examine the untapped potential for fertiliser use that still exists among the sample farmers. The untapped potential for fertiliser use is calibrated after deciphering (a) the proportion of non-growers of a crop among non-adopters of fertilisers, (b) growth-generating cultivators among the discontinuous users, and (c) relative importance of change in rates of fertiliser application and change in area fertilised of continuous users of fertilisers. Through these adjustments, the analysis in the section estimates the viable proportion of untapped potential existing on the sample farms.

Non-users of Fertilisers on a Crop

Tables 6.6 and 6.7 give some characteristics of non-users of fertilisers on different crops. Following noteworthy points emerge from these tables. First, although there are a sizeable number of non-users of fertiliser on different crops, all of them actually do not fall in the category of non-users. For, a proportion of these farmers actually do not grow the crop. For example, gram is not fertilised during all the years between 2001–02 and 2003–04 by 281 farmers, but in fact over 99 per cent of these non-users on gram have not grown gram in this period. Similarly, chillies during these years have not been fertilised by 251 farmers but about 99 per cent of these farmers have not grown chillies during these years. These finding indicates towards the fact that one need to take cognisance of non-growers of a crop before arriving at the untapped potential owing to non-users of fertilisers on different crops. Second, a corollary to the above finding is that across crops, there are few crops on which non-use of fertilisers is high. For rest of the crops, non-use of fertilisers is not a significant issue. In fact, crop-specific non-use reveals that 40 per cent of 162 non-users in jowar, 20 per cent of 226 non-users of bajra and 16 per cent of 273 non-users of fertilisers of kharif pulse constitute bulk of non-users in the sample. On other crops, non-users are between 1 and 12 per cent of the reported number of farmers not fertilising a crop in any of the 3 years between 2001–02 and 2003–04. Third, non-users on a crop are late adopters of fertilisers. Only 16 per cent of non-users on maize are using fertilisers for more than 15 years. On jowar, only 30 per cent of 162 non-users have adopted fertilisers over 15 years back, rest have adopted more recently. The figures for other crops also reveal similar patterns that depict that non-users on a crop are generally those who have adopted fertiliser use recently. Fourth, the farmers who do not apply fertiliser on a crop are not necessarily non-users. They have fertilised a number of crops (Table 6.7). The farmers who are non-users on maize had used fertilisers on other crops; 81 farmers who have not applied fertilisers on maize did use fertilisers on jowar, cotton and bajra and the average rate of fertiliser use between the period ranged from 16.2 kg to 23.6 kg per acre. Fifth, maize, cotton, jowar are the predominantly fertilised crops among non-users on specific crops. Per acre rate of fertiliser application of these farmers, even in 2003–04, ranges between 13 to 22 kg of NPK. The weighted average potential for fertiliser use is about 50 kg per acre. This implies that about 56 per cent to 74 per cent of additional fertilisers could be used on these farms.

Table 6.6
Non-Users of Fertilisers not Growing the Crop

Crop	*Number of farmers not fertilising crop in any of the 3 years*	*Percentage of cultivators not growing crop*			*Percentage of farmers who started use of fertiliser before 1988*
		2001–02	*2002–03*	*2003–04*	
Maize	81	92.6	92.6	92.6	16.1
Jowar	162	56.2	59.9	53.1	33.3
Cotton	131	97.7	96.9	95.4	19.1
Soyabean	237	89.6	88.2	88.6	27.0
Groundnut	246	91.5	89.8	91.5	28.9
Wheat	188	97.9	97.9	97.3	27.6
Gram	281	99.6	99.3	99.6	29.2
Chilli	251	99.2	99.2	98.8	27.5
Bajra	226	85.4	80.1	85.8	31.9
K. Pulses	273	83.9	84.2	84.6	29.3

Source: Based on primary data collected by the author.

Table 6.7
Other Main Crops Fertilised and NPK Use by Non-Users of Fertilisers on a Crop

Crop	*Main crops fertilised*			*NPK consumption in kilogram per acre*		
	2001–02	*2002–03*	*2003–04*	*2001–02*	*2002–03*	*2003–04*
Maize	J C B	J C B	J C B	16.21	23.59	20.82
Jowar	M C W	M C W	M C W	16.16	18.87	19.57
Cotton	M J W	M J W	M J W	12.08	13.16	12.97
Soyabean	M C J	M C J	M C J	16.23	18.14	18.52
Groundnut	M C J	M C J	M C J	17.52	19.64	19.17
Wheat	M C J	M C J	M C J	16.31	18.61	19.48
Gram	M C J	M C J	M C J	17.37	19.91	19.48
Chilli	M C J	M C J	M C J	15.69	16.82	16.84
Bajra	M C J	M C J	M C J	19.95	21.61	21.98
K. Pulses	M C J	M C J	M C J	18.72	19.92	20.73

M: maize; J: jowar; C: cotton; S: soyabean; G: groundnut; W: wheat; Gr: gram; Ch: chillies; B: bajra; P: kharif pulses.

But we need to calibrate this untapped potential of non-users (i) using the findings of Table 6.6 relating to percentage farmers not growing the crop, and (ii) recognising the fact that one need not accept non-users to adopt fertilisers on a crop at optimum rate. In fact, we should recognise that at the best non-users of fertilisers on a crop can be encouraged to use fertilisers at the rate that is being applied by continuous users. With these assumptions, the estimate of untapped potential shall be a viable estimate of additional fertiliser use if farmers are properly motivated by agricultural extension system. This exercise shall be attempted in the rest of the section. The analysis first tries to develop understanding on pattern of fertiliser use by infrequent users.

Infrequent User of Fertilisers on a Crop

Tables 6.8 and 6.9 give some characteristics of infrequent fertiliser users on a crop. This information pertains to sample cultivators who have applied fertilisers on a crop only one or two out of the three years between 2001–02 and 2003–04. Following points are worth noticing from these tables.

First, in comparison to non-users of fertilisers on a crop, infrequent users are relatively low. For example, there are 21 farmers in the sample who have applied fertilisers only one out of three years (Table 6.8) and only 15 sample farmers that have applied fertilisers on cotton only two out of the three years (Table 6.9). The combined number of infrequent users (applying two or less years between 2001–02 and 2003–04) on cotton is only 27 per cent of the number of farmers who do not apply fertilisers on cotton any of the three years. But when we account for non-cotton growing farmers, we find that non-users on cotton are only four while there are 27 infrequent users on cotton. Second, this implies that infrequent users could also be an important source of growth in fertiliser use; at least as important as the non-users. Subsequently, it shall be

Table 6.8
Infrequent (1 out of 3 Years) Users of Fertilisers not Growing the Crop

Crop	*Number of farmers fertilised crop 1 out of 3 years*	*Percentage of cultivators not growing crop*			*Percentage of farmers who started fertiliser use before 1988*
		2001–02	*2002–03*	*2003–04*	
Maize	11	54.5	54.5	27.3	9.1
Jowar	16	37.5	56.2	37.0	37.5
Cotton	21	76.2	57.1	57.1	19.0
Soyabean	9	55.6	33..3	—	33.3
Groundnut	17	41.2	47.1	58.8	29.4
Wheat	19	89.5	63.2	31.6	26.3
Gram	1	—	—	1	—
Chilli	5	—	—	—	20
Bajra	23	56.5	26.1	43.	39.1
K. Pulses	7	57.1	43.5	71.4	42.9

Table 6.9
Other Main Crops being Fertilised by Infrequent (1out of 3 Years) Fertiliser Users

Crop	*Main crops fertilised*			*NPK consumption in kilogram per acre*		
	2001–02	*2002–03*	*2003–04*	*2001–02*	*2002–03*	*2003–04*
Maize	C J W	C J W	C J W	12.36	35.46	14.53
Jowar	M W C	M W C	M W C	22.10	21.34	23.14
Cotton	M J W	M J W	M J W	11.11	15.13	19.36
Soyabean	M J C	M C J	M C W	20.21	21.53	22.83
Groundnut	M C S	M C J	M J W	14.97	18.10	15.71
Wheat	M C J	M C J	M C J	21.25	29.87	22.25
Gram	C W	C	M C W	0.19	0.35	0.23
Chilli	W M C	W M C	W M C	0.64	0.74	0.73
Bajra	M C J	M C J	M C J	1.87	2.14	2.20
K. Pulses	J M C	M C W	M C S	1.65	2.53	1.74

shown that this is partially true for some of the crops (Table 6.18). Third, like non-users of fertilisers, the infrequent users grow and fertilise a number of crops (Tables 6.9 and 6.11). But crop fertilised among the infrequent users varies significantly when compared to crop fertilised by non-users (Table 6.7). Apart from maize, cotton and jowar a significantly large number of other crops like wheat, chillies and soyabean are fertilised by the infrequent users. Fourth, it is also observed that there are few crops like groundnut and chillies as well as maize and soyabean where a sizeable proportion of sample cultivators who had used fertilisers for two out of three years have adopted fertiliser use since over 15 years. This proportion in the case of non-users on a crop (Table 6.6) is relatively low for all crops except wheat, bajra and kharif pulses. This implies that with longer experience of using fertilisers, the likelihood of being an infrequent user is more than that of being a non-user. Lastly, the average use on all crops taken together of infrequent users of fertilisers is relatively higher on growth-generating crops like maize, cotton and chillies. This further substantiates the fact that motivating infrequent users to continuously use fertilisers may add to future growth of fertiliser use as much as by motivating non-users to use fertilisers on a crop.

There is, however, internal fallacy with the argument; motivating all infrequent fertiliser users to use fertilisers regularly may not always result in growth of fertiliser use. For, those fertiliser users on a crop who are using them either one year or two years in which 2003–04 is included are already in the fold of fertiliser users. Therefore, only infrequent users who are non-users in 2003–04 are farmers who shall generate growth in fertiliser use in the future, if motivated to do so. In fact, (a) all those users who used fertiliser only one year either 2001–02 or 2002–03, and (b) all those farmers who used fertilisers for two years 2001–02 and 2002–03 are potential growth

Table 6.10
Infrequent (2 out of 3 Years) Users of Fertilisers not Growing the Crop

Crop	*Number of farmers fertilising crop 2 out of 3 year*	*Percentage of cultivators*			*Percentage of farmers who started fertiliser use before 1988*
		2001–02	*2002–03*	*2003–04*	
Maize	21	10.5	15.8	28.6	38.1
Jowar	27	—	18.5	22.2	25.9
Cotton	15	26.7	13.3	33.3	26.7
Soyabean	8	50	25	—	37.5
Groundnut	4	50	—	25	50
Wheat	7	28.6	28.6	0	14.3
Gram	—	—	—	—	—
Chilli	4	—	—	—	50
Bajra	6	16.7	16.7	—	—
Pulses	1	—	—	—	—

Table 6.11
Fertilised Crops and NPK use of Infrequent Users of Fertilisers (2 out of 3 Years)

Crop	*Main crops fertilised*			*NPK consumption in kilogram per acre*		
	2001–02	*2002–03*	*2003–04*	*2001–02*	*2002–03*	*2003–04*
Maize	J C Ch	C J W	C J S	19.78	22.16	22.76
Jowar	C M S	C M W	C M S	17.64	25.96	18.73
Cotton	M J S	M J S	M J S	14.41	15.17	17.61
Soyabean	C M Ch	C M J	C J M	12.86	17.61	14.52
Groundnut	J M S	J M S	S M W	19.66	27.49	34.56
Wheat	C M J	C M J	C M J	13.91	20.32	16.28
Gram	—	—	—	—	—	—
Chilli	C M W	C M W	C M W	19.50	27.77	30.61
Bajra	C J W	C J M	C M J	10.27	9.42	8.28
Pulses	M W B	C W B	M W B	18.79	18.17	21.58

generators if motivated to use them regularly on the crop. Rest of the sample infrequent users are using fertilisers in 2003–04 and may not be a source of future growth if they keep on using fertiliser. This fact has been kept in mind while untapped potential for fertiliser use in the case of infrequent users has been estimated subsequently (Table 6.17).

Fertiliser Use by Continuous Users

Before analysing the untapped potential for fertiliser use among the continuous users, it will be of interest to understand the characteristics of the continuous users and the source of growth in their fertiliser use. This understanding shall give us insight in the proportion of untapped potential of fertiliser use that could become viable source for future growth in fertiliser use.

Tables 6.13 and 6.14 give some characteristics of continuous fertiliser users on a crop. As all the users have applied fertilisers continuously, there is none among them who did not grow the crop. Moreover, the experience of using fertiliser on a crop among continuous users reveals that on all the crops except bajra and kharif pulses a sizeable proportion of farmers adopted fertilisers over 15 years back. The proportion of experienced fertiliser users among continuous users is significantly higher compared to either non-users or infrequent users on a crop; the crop-specific comparison reveals that on maize, cotton, soyabean, groundnut, wheat, gram and chillies the proportion of farmers using fertiliser for more than 15 years is significantly large for continuous users compared to non-users as well as those who applied fertilisers for only one year on the crop. A longer experience of fertiliser use gave confidence to farmers to use fertilisers continuously on a crop. The growth in fertiliser use of continuous users also reveals that between 2001–02 and 2002–03 all crops registered an increase in use except for soyabean (Table 6.15); soyabean registered a

Table 6.12
Continuous Users of Fertilisers not Growing the Crop

Crops	*Number of farmers fertilised crop all the 3 years*	*Percentage of cultivators not growing the crop*			*Percentage of farmers who started use of fertilisers before 1988*
		2001–02	*2002–03*	*2003–04*	
Maize	170	—	—	—	35.9
Jowar	78	—	—	—	20.5
Cotton	116	—	—	—	43.1
Soyabean	29	—	—	—	44.8
Groundnut	16	—	—	—	31.25
Wheat	69	—	—	—	36.2
Gram	1	—	—	—	100
Chilli	23	—	—	—	47.8
Bajra	28	—	—	—	7.1
Pulses	2	—	—	—	—

Table 6.13
Other Main Crops being Fertilised by Continuous Fertiliser Users

Crops	*Main crops fertilised*			*Fertiliser consumption in NPK kg per acre*		
	2001–02	*2002–03*	*2003–04*	*2001 –02*	*2002– 03*	*2003– 04*
Maize	C W J	C W J	C W J	20.55	20.44	21.60
Jowar	M C W	M C W	M C W	23.69	22.30	24.10
Cotton	M W J	M W J	M W J	24.29	11.56	26.26
Soyabean	M J W	M J W	M J W	31.07	32.39	32.48
Groundnut	M J C	J M C	M C W	35.98	34.14	36.33
Wheat	M C J	M C J	M C J	0.60	0.60	23.89
Gram	All (-B P)*	All (-B P)	All (-B P)	51.35	46.61	57.70
Chilli	C M J	C M W	C M W	29.48	33.01	32.81
Bajra	J M W	J M W	J M W	14.37	18.79	13.18
Pulses	All (–S G Gr)*	All (S G Gr)	All (-S G Gr)	28.02	38.52	32.55

* Except for crop on parentheses

decline in NPK use by about 628 kg. However, between 2002–03 and 2003–04, all crops except soyabean and chillies registered a decline in NPK use. This decline in use seems to be weather induced. Notwithstanding the decline, the important question is what is the main source of growth in the NPK use; is it change in rate of NPK use per acre that has driven this growth or is it the change in the area fertilised by the sample growers or both?

Table 6.14
NPK Use among Continuous Users

Crops	*No. of Farmers*	*Total NPK use in kilogram*			*NPK Use in kilogram per acre*		
		2001–02	*2002–03*	*2003–04*	*2001–02*	*2002–03*	*2003–04*
Maize	170	4443	4611	4567	21.9	22.4 (56)@	21.3
Jowar	78	1857	2030	1951	15.3	17.6 (65)	16.11
Cotton	116	7609	8939	8266	33.0	36.3 (40)	35.7
Soyabean	29	2026	1398	1973	30.9	21.2 (43)	31.1
Groundnut	16	494	510	503	21.2	21.9 (–)	21.7
Wheat	69	3391	3649	3535	31.4	35.1 (53)	32.9
Gram	1	216	216	216	27	27 (–)	27
Chilli	23	2406	3046	3171	55.6	64.8 (7)	72.49
Bajra	28	330	363	305	10.7	12.6 (75)	9.9
Pulses	2	15	34	30	18.6	43 (–)	37.4

@Figures in parentheses are untapped potential of continuous users on the crop.

Table 6.16 gives the relative importance of area fertilised and rate of NPK application per acre in the growth of NPK use of continuous users. What is evident from the Table 6.11 is that both growth in NPK use as well as deceleration in it amongst the continuous users on a crop is significantly governed by change in the rates of NPK use. The only exception of this are maize, cotton and chillies in which contribution of increase in area fertilised between 2001–02 and 2002–03 is also sizeable. Otherwise, on all the crops, growth as well as declaration in use has been significantly governed either by increase in rate or decline in it. Similarly, between 2002–03 and 2003–04 except soyabean and chillies all other crops registered a decline in total NPK use. Except groundnut and cotton, on all other crops the change in NPK use, in this period, was governed by change in rates of NPK use per acre. In the case of cotton and groundnut both fall in rates of application as well as fall in area fertilised between 2002–03 and 2003–04 are equally responsible for fall in total NPK use.

Table 6.15

Decomposition of Growth in NPK Use of Continuous Users (Kilogram)

Crop	*Between 2001-02 and 2002-03*				*Between 2002-03 and 2003-04*			
	$\Delta FA.R_O$	$\Delta R.FA_O$	$\Delta FA.\ \Delta R$	*Total*	$\Delta FA.R_O$	$\Delta R.FA_O$	$\Delta FA.\ \Delta R$	*Total*
Maize	51.81	115.30	1.34	168.46	192.44	–227.05	–9.48	–44.08
Jowar	–90.54	277.13	–13.51	173.08	98.43	–169.77	–8.23	–79.58
Cotton	519.22	759.27	51.81	1330.30	–538.40	–143.23	8.63	–673.00
Soyabean	9.27	–634.92	–2.90	–628.56	–53.45	654.02	–25.01	575.56
Groundnut	—	16.00	—	16.00	–2.19	–4.83	0.02	–7.00
Wheat	–122.58	394.03	–14.24	257.20	117.53	–224.50	–7.23	–114.20
Gram	—	—	—	—	—	—	—	—
Chilli	208.61	396.80	34.40	639.82	–210.61	361.36	–24.99	125.76
Bajra	–20.34	56.73	–3.50	32.89	23.84	–76.97	–5.06	–58.19
Pulses	—	19.50	—	19.50	—	–4.50	—	–4.50

The above findings clearly reveal that the engine of growth of fertiliser use of continuous users is increase in rate of application of fertiliser use while the contribution in increase in fertilised area is marginal. This is mainly because farmers are using fertilisers on nearly the entire cropped area. The future growth of continuous users shall, thus, be governed significantly by increasing rates of application. But increasing rates shall also be subjected to the ceiling forced by optimal rates as recommended by the agricultural scientists.

Table 6.16
Relative Importance of Various Sources of Growth in NPK Use of Continuous Users (Percentage)

Crop	*Between 2001–02 and 2002–03*				*Between 2002–03 and 2003–04*			
	Area effect	*Rate effect*	*Simultaneous effect*	*Total*	*Area effect*	*Rate effect*	*Simultaneous effect*	*Total*
Maize	30.8	68.4	0.8	100	–436.6	515.1	21.5	100
Jowar	–52.3	160.1	–7.8	100	–123.7	213.4	10.3	100
Cotton	39.0	57.1	3.9	100	80.0	21.3	–1.3	100
Soyabean	–1.5	101.0	0.5	100	–9.3	113.6	–4.3	100
Groundnut	0	100	0	100	31.3	69.0	–0.3	100
Wheat	–47.7	153.2	–5.5	100	–102.9	196.6	6.3	100
Gram	—	—	—	100	—	—	—	100
Chilli	32.6	62.0	5.4	100	–167.5	287.4	–19.9	100
Bajra	–61.9	172.5	–10.6	100	–41.0	132.3	8.7	100
Pulses	—	100	—	100	—	100	—	100

Future Growth in Fertiliser Use

The above analysis provided us with a number of valuable insights with respect to pattern of fertiliser use. In order to estimate future growth and its sources, the analysis also provided understandings that can be used to develop conditions governing the exploitation of untapped potential for fertiliser use. Table 6.17 gives the key indicators used to estimate untapped potential for fertiliser use of (a) non-users on a crop, (b) infrequent users on a crop, and (c) continuous users of fertilisers on a crop.

The optimum fertiliser use based on recommendation as well as discussion with extension officers of the area is given for all the crops in Table 6.17. The potential for additional use of NPK for each of the crops in the case of non-users on the crop is based on (a) number of farmers growing a crop but not fertilising it in any of the three years, (b) the area remained unfertilised on the crop, (c) the optimum rate, and (d) the rates of application of continuous users. The potential for additional use of NPK for each of the crops in the case of infrequent users is based on the level of use of (a) only 2001–02 users, (b) only 2002–03 users, and (c) only 2001–02 and 2002–03 users on the crop. The potential for additional use of NPK for each of the crops in the case of continuous users on the crop is based on (a) proportion of untapped potential and (b) level of NPK use of continuous users (Table 6.17).

A cursory glance on Table 6.17 reveals that although a significant area under jowar, kharif pulses, soyabean and bajra has remained unfertilised, the level of untapped potential is relatively low in bajra and jowar. Consequently, the most viable and sizeable untapped potential for fertiliser use shall be only on jowar and soyabean. For other crops, either area unfertilised of non-users for all the three years on a crop is very low or the proportion of untapped potential is insignificant. Both these conditions would make the untapped potential low as well as difficult to harness. On the other hand, the proportion of untapped optimum rate for kharif pulses, gram and groundnut has already been exhausted. Therefore, the future use of continuous years on these crops shall be nil or very low. In the case of chillies too, as the level of untapped potential is low (only 7%), future growth in NPK use through improving rates of application may not be significant and viable to exploit.

Table 6.17
Parameters for Estimating Untapped Potential of NPK Use

Crop	*Non-users*			*Infrequent users*[1]	*Continuous users*		
	Number of farmers	*Unfertilised area in acres*	*Proportion of optimal rate tapped*		*NPK use*	*Proportion of optimal rate untapped*	*Optimal rates*
Maize	6	5	0.44	276	5318	0.56	50
Jowar	65	99	0.35	298	2690	0.65	50
Cotton	4	4	0.60	189	9461	0.40	60
Soyabean	28	40	0.57	74	1549	0.43	37
Groundnut	25	30	1.00	116	750	0	15
Wheat	4	2	0.47	176	4203	0.53	75
Gram	2	1	1.00	–	436	0	15
Chilli	2	2	0.97	20	3260	0.07	70
Bajra	45	40	0.25	84	609	0.75	50
Pulses	43	43	1.00	16	101	0	25

1. Based on Table 6.3.

Using parameters of Table 6.17, one can estimate the untapped potential under different patterns of fertiliser users. For non-users for each crop, the untapped potential (U_{NU}) can be estimated as follows:

$$U_{NU} = \text{area unfertilised} \times \text{proportion of optimal rates tapped} \times \text{optimum rate.}$$

The untapped potential for infrequent users (U_{IU}) each crop is:

$$U_{IU} = \sum_{t=1}^{3} NPK_t$$

where, NPK_1 and NPK_2 are respectively fertiliser nutrient applied by 2001–02 users who discontinued use during 2002–03 and 2002–03 fertiliser users who discontinued use during 2003–04. NPK_3 is two-year users who discontinued use during 2003–04.

The untapped potential for continuous users of each crop (U_{CU}) can be estimated as follows:

$$U_{CU} = (\text{level of NPK/proportion of optimal rate untapped}) \times \text{untapped rate}$$

Adding U_{NU} or U_{IU} or U_{CU} for all crops would give the total untapped potential for non-users, infrequent users and continuous users respectively. These estimates are presented in Table 6.18. Findings of Table 6.18 are not very different from the expected; about 32,400 kg of additional NPK use can be expected in near future if extension system tries to increase the rate of fertiliser use. This expected addition use is about 49 per cent of the total potential for NPK use of the sample farmers. Put other way, it can be concluded that about 80 per cent of total potential is remaining untapped owing to inefficient use by continuous users, 16 per cent remains untapped owing to non-users and 4 per cent owing to infrequent users who dropout of fertiliser use frequently. It is interesting to note that in low fertiliser using areas like western tribal belt of Madhya

Pradesh, non-adoption of fertilisers, non-use of fertilisers on a crop and infrequent fertiliser use is not a serious problem. Bulk of the inefficiency is owing to fertiliser use at relatively low rates.

Table 6.18

Untapped Potential for Fertiliser Use for the Sample Farmers

	Non-users	*Infrequent users*	*Continuous users*
Maize	110	276	6768
Jowar	1732	298	4995
Cotton	144	189	6307
Soyabean	844	74	1168
Groundnut	450	116	—
Wheat	71	176	4739
Gram	15	—	—
Chilli	130	20	245
Bajra	500	84	1827
Pulses	1075	16	—
All Crops	5071 (16)	1249 (4)	26049 (80)

The extension agency need not become highly diffused in improving efficiency of the fertiliser use in the area, for findings of Table 7.13 reveal that concentrating on three crops will result in achieving over 70 per cent of untapped potential for fertiliser use. Non-users of jowar, kharif pulses and soyabean if start using fertilisers on these crops, would account for over 70 per cent of growth in fertiliser use that could be achieved if all the non-users are encouraged to start using fertilisers. It implies that extension system can become selective in propagating fertiliser use in tribal areas. They could start with jowar, kharif pulses and soyabean and over time move to other crops. If infrequent users of jowar, maize and cotton are encouraged to use fertilisers regularly, the additional growth by these farmers shall be over 60 per cent of total growth that could be created due to regular use by infrequent users. On the other hand, the growth of NPK use if continuous users of maize, cotton, jowar and wheat increase their rates of application would be over 88 per cent of total growth due to improvement in rates on all crops. This also implies that improving the efficiency of fertiliser use on cotton, maize, jowar and wheat is the key to achieving over 80 per cent of untapped potential. The task of the extension system is now much easier.

How this inefficiency can be minimised? Alternatively, how the growth in fertiliser use can be sustained? A portion of this inefficiency is on account of farmers' perceptions that optimum use on a crop is actually lower than what is recommended by the extension agency, even under very difficult conditions of remote tribal areas. But a part of the inefficiency could also be owing to variations of the conditions that influence fertiliser use; the farm conditions. Some of these variations shall be analysed in terms of their influence on pattern of fertiliser use on farms in the next chapter.

V

Summing up

Fertiliser use of the sample farmers has grown from 24,800 kg in 2001–02 to 28,400 kg in 2002–03. The use thereafter stagnated at 28,100 kg in 2003–04. The sluggishness in use in 2003–04 has been an outcome of fall in use in all the crops, except chillies and soyabean. Despite the sluggish growth and fall in NPK use, three crops—cotton, wheat and jowar—account for two-thirds of total use of NPK. In this, cotton alone accounts for about a third of total use. Despite stagnation and fall in NPK use during 2003–04, the use in the year is still higher in comparison to 2001–02. The increase in NPK use between 2001–02 and 2003–04, on three crops—cotton, chillies and wheat—accounts for about two-thirds of growth in fertiliser use. Groundnut and kharif pulses had shown a decline in fertiliser consumption while growth in fertiliser use on gram was nil.

The contribution of different sources of growth in fertiliser use has been attempted for crops that have shown substantive growth, moderate growth and negative growth. Following points emerge from the analysis:

- Growth pattern reveals that in crops like wheat, chillies and cotton, total increase in NPK use was substantial; in crops like maize, jowar, soyabean and bajra the growth in NPK use was moderate while in the case of pulses, groundnut and gram the growth in NPK use was negative or stagnant.
- Farmers continuously applying fertilisers during all the three years contribute to total growth in NPK (a) substantially in the case of chillies and pulses, (b) moderately in the case of maize and jowar, (c) low in the case of groundnut, and (d) negative in the case of bajra and soyabean.
- A substantial reduction in growth in fertiliser use has taken place owing to discontinuation of use in the case of maize and bajra, during 2003–04 and in the case of jowar, pulse and groundnut for both 2002–03 and 2003–04.
- The sources of growth vary significantly across crops. (a) For wheat it is new adopters that created substantial growth in NPK use while in the case of cotton and chillies it is continuous users as well as new adopters that generated growth in NPK use; (b) in the case of maize, jowar, soyabean and bajra, the new adopters did substantially contribute to growth in NPK use; and (c) in the case of low growth crops kharif pulses, groundnut and gram, low contribution of continuous users as well as cancelling out of effect of drop-out and new adopters resulted in low growth in NPK use.

A large untapped potential in wheat and cotton, has, thus, been owing to discontinuation of use. In the moderate growth group, maize, jowar, soyabean and bajra, the untapped potential is significantly high owing not only to discontinuous use but also owing to low use by continuous users. The negative growth of crops pulses, groundnut and gram has demonstrated not only low growth in the use by continuous users but also a high discontinuation of use. In this group of crops, not only adoption is low and the continuous use is contributing marginally but also the discontinuous use is very high. In fact, the use pattern of pulse, gram and groundnut shows, a high untapped potential at different levels.

The analysis in this chapter reveals that about 3,24,000 kg of additional NPK use can be expected in near future if extension system tries to minimise the inefficiency in fertiliser use. About 80 per cent of total potential is remaining untapped owing to inefficient use by continuous users, 16 per cent remains untapped owing to non-users and 4 per cent owing to infrequent users who dropout of fertiliser use frequently. This expected addition to fertiliser use is about 49 per cent of the total potential for NPK use of the sample farmers. In low fertiliser using areas like western tribal belt of Madhya Pradesh, non-adoption of fertilisers, non-use of fertilisers on a crop and infrequent fertiliser use is not a serious problem. Bulk of the inefficiency is owing to fertiliser use at relatively low rates.

The extension agency need not become highly diffused in improving efficiency of the fertiliser use in the area because our findings indicate that:

1. Concentrating on three crops where non-use is high will result in achieving over 70 per cent of untapped potential for fertiliser use. Non-users of jowar, kharif pulses and soyabean if start using fertiliser on these crops, would account for over 70 per cent of growth in fertiliser use that could be achieved if all the non-users are encouraged to start using fertiliser. It implies that extension system can become selective in propagating fertiliser use in tribal areas. They could start with jowar, kharif pulses and soyabean and over time move to other crops.
2. If infrequent users of jowar, maize and cotton are encouraged to use fertilisers regularly, the additional

growth by these farmers shall be over 60 per cent of total growth that could be created due to regular use by infrequent users.

3. The growth of NPK use if continuous users of maize, cotton, jowar and wheat increase their rates of application would be over 88 per cent of total growth due to improvement in rates on all crops.
4. This also implies that improving the efficiency of fertiliser use on cotton, maize, jowar and wheat is the key to achieving over 80 per cent of untapped potential existing. The task of the extension system is now much easier.

A portion of the inefficiency is on account of farmers' perceptions that optimum use on a crop is actually lower than what is recommended by the extension agency, even under very difficult conditions of remote tribal areas. But a part of the inefficiency could also be owing to variations of the conditions that influence fertiliser use, the farm conditions. The next chapter deals with these.

Chapter 7

Explaining Pattern of Fertiliser Use at Farm Level

Findings of the preceding chapters reveal that pattern of fertiliser use not only varies across crop but in a crop there are significant variations; some crops like maize, jowar, cotton and wheat have attracted more farmers for early adoption of fertiliser use. On some crops, the diffusion of fertiliser use to other crops is much faster. It was also observed that fertiliser use pattern has a peculiarity across farms; some farmers have yet to adopt fertiliser use while some use fertilisers infrequently and some are continuous users. The objective of this chapter is to identify determinants of the pattern of fertiliser use at farm level. It can be argued that a portion of inefficiency, in the form of non-adoption, non-use and discontinuous use, is on account of variation in farm conditions. Two sets of variables have been identified that could have significant association with the observed pattern of fertiliser use at farm level: the farm characteristics that are internal to the farmer. The important indicators in the group are caste status, family size and structure, education level; size of holding and number of fragmentation, tenancy status, soil type and topography of plot; income of the household, level of irrigation and relative importance of agriculture in total income of farmer. The second set of indicators that could have significant association with the pattern of fertiliser use could be external support available for rapid adoption and diffusion of fertiliser use. These supports could be in the form of extension support, fertiliser use techniques, institutional credit availability and input-out marketing.

Technology and Market Access

The western tribal belt is located in a harsh agro-economic environment. In the last fifteen years, agricultural practices have changed as energised irrigation has helped make seed-fertiliser technology transfer possible in about 15 per cent of the area. Farmers have been incorporated into local and national markets, and farm households have moved to high external input agriculture. In addition, as cotton, wheat and soyabean have entered farming systems, the proportion of production for subsistence has declined. Combined, these innovations have increased the monetisation of the local economy and its engagement in the wider cash economy. The unavailability of cash saving opportunities and the almost entire absence of non-farm enterprise in the area results in the need to borrow seasonally—for seed, agro-chemicals and periodically for capital investments for the purchase of irrigation pumps, investment in digging irrigation wells and for contingencies such as bride price payments and funeral expenses. The villagers in the area do not have easy access to formal banking. Banias in nearby markets are the sole source of credit and have a monopolistic position, charging high rates of interest.

Despite the fact that farmers' perception about response of fertiliser use has been governed by this environment. In this environment they are not very wrong when they perceive that improved varieties respond better than HYVs, and good rainfall is much more motivating to increase fertiliser use than irrigation. But the diffusion of fertiliser use has taken place under poor technology environment whereby the crop-specific agronomic practices and fertiliser use technology is in variation with recommendations of the state extension agency. A large number of farmers were aware about chemical fertilisers since they were introduced to agriculture by their elders. But still, the actual use started late. About 94 per cent of farmers use fertilisers. Nonetheless, three important factors that determine quantity used by farmers are rainfall, availability of irrigation and availability of short-term credit for input purchase. Farmers also reported that their first priority is to apply fertilisers on relatively plain plots where soils is deep, especially those patches that could retain moisture relatively for a large period. A minuscule of wheat growers (only 16%) and maize growers (5%) have heard about optimal recommendations of fertiliser use. Although, only 12 per cent farmers are aware about sources of various nutrients, they have relatively more knowledge about use of fertiliser materials urea, DAP and SSP in basal and top dressing. Over 58 per cent farmers know that urea can be used both in top and basal dressings, and about half of the cultivators are aware that DAP and SSP should not be used in top dressing. On the other hand, only 2 per cent farmers are aware about micronutrients and one out of ten farmers knows about soil test. Half of those who have known soil test, have had ever gotten their soil tested and a minuscule of them received the result of soil test.

Nearly the entire marketed surplus finds a market just after harvest. Only 15 per cent cultivators have reported that they store their output for a better price. Over two-third marketed surpluses find its way to either the bania or village trader, only a few prefer to go to agricultural market yard or the cooperative market channel. Outlets like bania and nearby markets also dominate the input purchase. Only 40 per cent of the sample fertiliser users had preferred purchase from cooperative channels. Less than a third of sample farmers are members of cooperative society, and only two out of ten farmers have received loan for fertilisers from village cooperatives. Long history of defaults in repayment and non-transferor of agricultural land in their names are two major reasons that have diverted farmers for fertiliser purchase from private trade. A third of farmers reported that the private loans for inputs were received after mortgage but over 60 per cent of those farmers who had taken loans reported that they were able to service their non-institutional credit.

A wide spread fertiliser use but at low and inefficient way forces us to construct the technology index with limited data. The analysis in this chapter has used indices for extension support, fertiliser use technology, access to institutional finance and input-output marketing. But as technical knowledge about fertilisers and other input use is poor and developed without extension support, the technology status and farm practices do not significantly vary across the sample. As a result, in construction of these indices, use of actual farm practices has not received due weight. The following method has been adopted in construction of the four indices that are used in subsequent analysis.

Fertiliser Use Technology (FUT) Index

As farmers' knowledge about time and methods of fertiliser use widely vary from the recommendations, instead of evaluating levels of fertiliser use and methods of application, the FUT Index is developed aggregating the following indicators:

If sources of N are spelt correctly	1, 0 otherwise
If sources of P are spelt correctly	1, 0 otherwise
If sources of K are spelt correctly	1, 0 otherwise
If method of basal use of P is spelt correctly	1, 0 otherwise
If method of basal use of K is spelt correctly	1, 0 otherwise

If method of Top dressing of N is spelt correctly	1, 0 otherwise
If Soil test is done	1, 0 otherwise

The FUT Index ranges from 0 to 7.

The Extension Support (ES) Index

The ES Index that ranges from 0 to 5 is developed using the following variables:

If received support from VLW	1, 0 otherwise
If contact farmers of the extension agency	1, 0 otherwise
If known to the extension office	1, 0 otherwise
If met the officer in last 3 months	1, 0 otherwise
If used the recommendation	1, 0 otherwise

Input Output Marketing (IOM) Index

Following variables were used to calculate IOM Index:

If output is sold at market yard or cooperative institutions	1, 0 otherwise
If sales were made at peak prices	1, 0 otherwise
If inputs were purchased from institutions	1, 0 otherwise

The IOM index thus ranges from 0 to 5.

Credit Support (CS) Index

The CS Index, which ranges from 0 to 3, has been calculated using following information

If member of village cooperative	1, 0 otherwise
If non-institution debts are serviced	1, 0 otherwise
If repayment is done by income from agriculture	1, 0 otherwise

Explaining Pattern of Fertiliser Use

An analytical approach for evaluating the association between different factors and the pattern of fertiliser use could be two-way classification between the two. The findings of the two-way classification can further be substantiated by the significance of chi-square value and the correlation between these. Table 7.1 gives the classification of the sample farmers across size class of their land holdings and their pattern of fertiliser use. A cursory glance on the table reveals that relatively larger proportion of small cultivators is in the category of non-adopters; this proportion reduces as one moves from small to large farms. On the other hand, the proportion of large farmers using fertilisers continuously is relatively larger compared to this proportion in medium and small farms. All this implies that there may be a positive association between size of farm and the pattern of fertiliser use; infrequent fertiliser use is relatively more on small farms whereas on large farms, fertiliser use is relatively continuous.

The visual observation of the Table 7.1 could be substantiated by the significance of chi-square and the sign and significance of the correlation co-efficient between the pattern of fertiliser use and size of landholding. While the value of chi-square is 14.5 and its significance is 0.07, the correlation co-efficient between the two is +0.17 with significance level 0.003. Together these statistics reveal that (i) there is a strong

Table 7.1

Association between Size of Landholding and Pattern of Fertiliser Use

Pattern	*Landholding size*			
	Small	*Medium*	*Large*	*All*
Non-adopters	8	7	3	18
Non-users	2	2	3	7
Used only for 1 year	3	2	2	7
Used for 2 years	5	8	1	14
Continuous users	54	94	107	255
All	72	113	116	301

Source: Based on primary data collected by the author.

association between the two variables and (ii) on farms with relatively larger landholding, the pattern of fertiliser use is relatively more continuous than otherwise. To identify the association between different other factors with pattern of fertiliser use, a similar exercise is done using all social, economic and support factors. Findings of the analysis are presented in Table 7.2.

Among social factors, none other than education is associated with the pattern of fertiliser use. It means that whether one is tribal or from Scheduled Caste or Patidar, there is no significant variation in the pattern of fertiliser use of the farmers. Size of family and the family structure are also not associated with the pattern of fertiliser use. But among literates, a significantly larger proportion of farmers are continuous users rather than discontinuous users, whereas among illiterates, significantly larger proportion of households are non-users or infrequent users.

Among economic variable fragmentation of land holding and tenancy has no association with pattern of fertiliser use. On the other hand, longer experience of fertiliser use, larger size of holdings, farms with deep soil, farms with large proportion of area being plain, high income households, households where agriculture is substantially contributing to family income and farms with larger proportion of area under irrigation have relatively more continuous fertiliser users. While, those with few years' experience of fertiliser use, small size of holdings, rocky soils with larger proportion of area undulating, low income households with larger proportion of income originating from labour and migration, and farms with no irrigation facilities have relatively more non-users and infrequent users of fertilisers.

More importantly, two support factors: Fertiliser Use Technology (FUT) Index and Input-Output Market (I-OM) Index are significantly and positively associated with the pattern of fertiliser use. But credit support and extension support are not able to influence significantly the pattern of

Table 7.2
Chi-square and Correlation between Pattern of Fertiliser Use and Explanatory Factors

Explanatory factors and their category	*Chi-square*	*Correlation*
Socio-Economic		
Ethnicity (Tribe, SC to Patidar)	4.0 (0.86)[a]	0.03 (0.58)
Size of Family (Small to Large)	4.7 (0.78)	0.04 (0.47)
Structure of Family (Nuclear to joint)	2.7 (0.61)	0.04 (0.45)
Education (Illiterate to Literate)	8.4 (0.07)	0.16 (0.01)
Economic		
Experience of Fertiliser Use (New adopters to experienced)	322.0 (0.00)	0.50 (0.00)
Size of Land holding (Small to Large)	14.5 (0.07)	0.17 (0.00)
Number Fragmented Plots (Few to Large)	15.5 (0.11)	0.07 (0.22)
Tenancy Status (None to Tenancy)	6.0 (0.20)	–0.08 (0.16)
Net Short-Term Land Transactions*	8.3 (0.40)	–0.11 (0.05)
Soil Type (Rocky to Deep soil)	16.3 (0.04)	0.19 (0.00)
Plain Area (Nil to High proportion)	17.1 (0.03)	0.12 (0.04)
Income of household (Low to High)	20.9 (0.01)	0.25 (0.00)
% of Agriculture in Total Income (Low to High)	21.2 (0.01)	0.23 (0.00)
Level of Irrigation (Nil to High)	20.2 (0.01)	0.22 (0.00)
External Support		
Fertiliser Use Index (Low to High)	48.4 (0.01)	0.30 (0.00)
Extension Support Index (Low to High)	27.6 (0.12)	–0.06 (0.27)
Credit Support Index (Low to High)	17.7 (0.12)	0.19 (0.00)
Input-Output Market Index (Low to High)	38.9 (0.00)	0.27 (0.00)

a Figures in parentheses are level of significance.

* Leased out to leased in.

fertiliser use. This is a reflection of poor access to institutional finance and poor quality of extension support in the area. But interestingly all those who know more about the sources of fertiliser nutrients and the time and method of use of these nutrients as well as better access to institutional marketing channels for inputs and output are relatively more continuous users.

What the above findings clearly reveal is, like adoption of fertilisers, its pattern of use too is governed by farm characteristics rather than farmer's characteristics. Improving (a) the soil-moisture of farm, (b) ways to apply fertilisers, and (c) access to institutional markets for input-output, are far more important in changing a discontinuous use pattern towards a continuous use pattern. Accumulated experience is just not the number of years since one is using fertilisers. It is also change in attitude towards fertiliser use, knowing the nuances associated with its use and experiencing productivity gains due to fertiliser use. The need, therefore, is to calibrate the extension messages and access to credit on the one hand and farm productivity gains through public investment on the other.

Inefficiency in Fertiliser Use

The last section has tried to understand why on some farms, the pattern of fertiliser use is continuous and on some the pattern is of infrequent use. Equally important is the inefficiency of fertiliser use that constraints optimum use of fertiliser. Table 7.3 gives some clues as to what the farmers believe constraints additional use of fertiliser on their farms. Interestingly, over a third of the respondents believe that soil-moisture constraints do not allow them to increase their fertiliser use. But over half of the farmers either show lack of extension support in the form of know-how of fertiliser use or there is utter confusion relating to their perceptions about fertiliser use. The misconceptions range from quality and fertility of soil deteriorates with fertiliser use or pest attack increases after fertiliser use. Consequently, these farmers prefer to use fertilisers at relatively low rates.

Table 7.3
Farmers' Beliefs about Constraining Factors for Additional Fertiliser Use

Factor	*Percentage*
Soil moisture constraints	38
Poor methods of fertiliser use	30
Perception inconsistency	22
No serious problems as rates are low	10
	100 (124)*

*Number of farmers accepting that there is a problem in added fertiliser use; another 125 farmers did not have any problem.

The perceptive incongruence in fertiliser use between farmers' beliefs and the recommendation of extension system are reflected when yield-response under changing price of fertiliser were analysed in the case of 121 wheat growing farmers. A hypothetical situation was created whereby these wheat growers were asked about their first reaction to a 10 per cent increase in urea prices. Two-thirds of the wheat growers responded with no change in their rates of application. But a third of wheat growers reported that they would reduce their fertiliser use on the crop. This reduction in fertiliser use was ranging from 10 to 55 per cent, with an average reduction in rates of about 20 per cent. They also believed that such reduction in fertiliser use would reduce their wheat yield by about 25 per cent, a range of 10 to 50 per cent reduction in wheat yield. The most expected response to the question if there is net loss in wheat cultivation due to reduced fertiliser use, would they like to increase their rates of fertiliser use, was yes. But in the sample, over two-thirds of wheat growers whose first reaction was to reduce rates, gave their second reaction that they would still continue using about 20 per cent less fertiliser, despite a net loss. Only a third of wheat growers whose first reaction was a reduction in use, as their second reaction reported getting back to old levels in order to avoid losses.

The preceding analysis indicates that farmers' short-term reaction to price rise was rational but the long-term reaction for a subset was irrational. Even considering the harsh agro-economic environment in which they have to make these decisions, a second reaction of this kind is, at the most, emotional. But what this clearly brings out is the need for a rational extension message suited for the tribal environment. It also underlines the need for widespread on-farm experiments to demonstrate the optimality in fertiliser use under varying environments. This tribal area represents relatively poor endowment and still poor agricultural infrastructure. All the major crops like maize, jowar, pulses are grown largely under unirrigated conditions. Some irrigation that exists is applied for long duration crops like cotton or rabi cultivation, mainly wheat. Except for pulses and groundnut, the level of fertiliser use is generally lower by about 60 per cent of the recommended rates for different crops.

At very low rates, a rise in fertiliser prices may not result in fall in fertiliser consumption. This is owing to the fact that the optimum rates in the new changed price environment may still be higher than the present rates of fertiliser use. This is, in fact, the situation in the western tribal belt. Fertiliser price rise for this group of farmers is a non-issue. The fertiliser marketing environment—whereby urea monopoly sale through cooperatives has broken—has set in at a time when the farmers' economy has faced serious problem due to recurring monsoon failure. As a result, large number of farmers have chosen to raise price issue to voice their grievances against the neglect of the area, especially with respect to poor agricultural infrastructure, high risk in adopting available technology, low productivity and high irrigation cost. The move to either decontrol fertiliser prices or decontrol marketing of urea is being read as a move to further marginalise the cultivators in the tribal region. Although our field investigation suggests that the farmers' reaction to price rise is mixed, findings also suggest that farmers' attitudes do change as more and relevant information is made available to them. The marketing network and state extension system could play a role to shorten the time lag in adjustment in rates of fertiliser application.

Independent of fertiliser prices, a number of other and equally important issues are thrown up by our investigation. Forces like lack of irrigation facilities and consecutive crop failure, poor soil with low agricultural productivity have resulted in creating perceptions that low rates of fertiliser use is a better risk management device in uncertain yield-response environment. In the short run, therefore, perceptions about yield-response to added fertilisers should significantly improve. Tribal farmers must receive accurate information about implications of changes in price and non-price factors on returns under changing operational environment of the region. Their understanding at present is being shaped under limited information. Consequently, a large section of wheat, maize, jowar and cotton growers in the area are operating at sub-optimal levels. Both low rates and low technical efficiency of NPK use is responsible for this. The watershed development in tribal areas has created environment for technology transfer at many locations. Such experiments by civil society organisation have been much more successful in improving efficiency of tribal agriculture (Sah and Bhatt, 2004).

The messages, therefore, need to focus on proper understanding on the optimality in fertiliser use, response-functions and relative importance of non-price factors in determining the farm productivity. The next agenda, therefore, should include issues like fertilisers and water use efficiency, complementary use of phosphate and urea, chemical fertilisers and farm yard manure, use of micronutrients, and management of farm practices. Instead, a mechanical extension messages, changing farmers' attitudes through counselling are an unexplored option to use inputs, especially chemical fertilisers, in harsh environment. The message, as the findings of last chapter reveals, should be concentrated on cotton, maize, jowar and wheat.

Summing up

This chapter examined why on some farms fertiliser use was continuous, on some infrequent and on still some fertiliser use has yet to start. The findings of the chapter reveal that among social factors, none other than education is associated with the pattern of fertiliser use. But among literates, a significantly larger proportion of farmers are continuous users rather than discontinuous users, whereas among illiterates, significantly larger proportion of households are non-users and infrequent users. Among economic variable, fragmentation of land holdings and tenancy has no association with pattern of fertiliser use. On the other hand, longer experience of fertiliser use, larger size of holdings, farms with deep soils, farms with large proportion of farm area being plain, high income households, households where agriculture is substantially contributing to family income and farms with larger proportion of area under irrigation have relatively more continuous fertiliser users. More importantly, two support factors: fertiliser use technology and input-output market support are significantly and positively associated with the pattern of fertiliser use. But credit support and extension support are not able to influence significantly the pattern of fertiliser use. This, in a sense, reflects on poor access to institutional finance and poor quality of extension support in the area. Interestingly, all those who know more about the sources of fertiliser nutrients and the time and method of use of these nutrients as well as better access to institutional marketing channels for either inputs or output are relatively more continuous users.

What the above findings clearly reveal is, like adoption of fertilisers, its pattern of use too is governed by farm characteristics rather than farmer's characteristics. Improving (a) the soil-moisture of farm, (b) ways to apply fertilisers, and (c) access to institutional markets for input-output, are far more important in changing a discontinuous use pattern towards a continuous use pattern. The need, therefore, is to calibrate the extension messages and access to credit on the one hand and farm productivity gains through public investment on the other for improving the pattern of fertiliser use. Interestingly, over a third of the respondents believe that soil moisture constraints do not allow them to increase their fertiliser use. But over half of the farmers either show lack of extension support in the form of know-how of fertiliser use or there is utter confusion relating to their perceptions about fertiliser use. Consequently, these farmers prefer to use fertilisers at relatively low rates.

The analysis relating to effect of price increase of fertiliser on its use indicates that farmers' short-term reaction to price rise by a third of farmers to reduce the rate of application was rational but their long-term reaction that they would continue low fertiliser use despite a net loss due to yield reduction, for a subset, was irrational. What this clearly brings out is the need for a rational extension message, suited for the tribal environment. This tribal area represents relatively poor endowment and still poor agricultural infrastructure. All the major crops like maize, jowar, pulses are grown largely under unirrigated conditions. The level of fertiliser use is generally lower by about 60 per cent of the recommended rates for different crops. Despite the fact that fertiliser prices for this group of farmers is a non-issue, the changes in marketing environment has set in at a time when the farmers' economy has faced serious problem due to recurring monsoon failure. As a result, large number of farmers have chosen to raise price issue to voice their grievances against the neglect of the area, especially with respect to poor agricultural infrastructure, high risk associated with adoption of available technology, low productivity and high irrigation cost. Nonetheless, independent of fertiliser prices, a number of other and equally important issues are thrown up by our investigation. Forces like lack of irrigation facilities and consecutive crop failure, poor soil with low agricultural productivity have resulted in creating perceptions that low rates of fertiliser use is a better risk management device in uncertain yield-response environment.

Chapter 8

Summary, Conclusions and Policy Implications

A significant proportion of growth in fertiliser use in Madhya Pradesh has taken place in irrigated areas where the potential for further growth is low. Tribal belt, a less endowed region, with low irrigation and uncertain weather conditions, has also contributed to this growth process; the western tribal belt of Madhya Pradesh consumes as high as 60 kg of NPK per acre in comparison to about 95 kg in the heartland of Malwa region. But the high rates are confined to plain irrigated areas where dominant caste is Patidars. In remote tribal areas the use is a fourth of this. A large proportion of fertilisers in this tribal belt is consumed during kharif. Due to frequent crop failures the cropping pattern is quite diversified in this area. A high proportion of fertiliser consumed in the region is in the form of nitrogen. A large proportion of potential for fertiliser use in this area is still untapped. This area not only contributes to growth of fertiliser use but also is responsible for all the fluctuations and stagnation in it even during normal weather years.

The above discussion clearly brings out the contours of future growth in fertiliser use in Madhya Pradesh. Tribal belt in Madhya Pradesh is important not only for creating equilibrium of fertiliser use across geographical locations in the state but it is also the most likely source for the future growth. The present study is diagnostic in nature and focuses on (a) economic forces affecting the cultivators' adoption and diffusion process of fertiliser use, and (b) analysis of the dynamics of fertiliser use for three continuous period of time between 2001–02 and 2003–04. A number of processes could be responsible for lowering the growth in fertiliser use, the non-users could be the only one of the inefficiency. Equally important shall be identification of the inefficiency in fertiliser use caused by infrequent users and inefficiency in fertiliser use caused by continuous users. As non-price factors behind the growth process have been more important than price factors, the analysis has been taken up crop by crop rather at the farm level. It was recognised that forces behind the growth in fertiliser use cannot be deciphered correctly without taking into cognisance the macro policy environment in which fertiliser use is growing. This became all the more important, for both demand for and supply of fertilisers has remained under state control. The study attempts to understand the following three processes:

> First, understand the pattern of adoption of fertilisers on different crops in this tribal area and evaluate the time lag after adoption farmers did experience in diffusing fertilisers.
>
> Second, understand the patterns in fertiliser use on different crops for three years between 2001–02 and 2003–04.
>
> Thirdly, estimate and examine the quantum of most viable untapped potential for fertiliser use and factors explaining the patterns observed at the farm level.

Summary of Major Findings

Data and Methods

Fertiliser use is an outcome of two variables: proportion of area under different crops receiving fertilisers and the rate of fertiliser application. Since different crops respond differently to fertiliser use and also because of difference in the crop prices, a varying pattern of fertiliser use would emerge in different agro-climatic zones. Fertiliser use would increase even in comparatively unfavourable price environment provided (a) the increase in potential for fertiliser use due to technological changes more than offsets the adverse effect of change in price environment; (b) the marginal returns to move from low level towards optimum level of fertiliser use are economically attractive to cultivators; and (c) the cultivators' expectation regarding the physical productivity of fertiliser use has improved.

The review of available literature on economics of fertiliser use has provided enough insight on forces governing fertiliser use at micro level but failed to explain the growth of fertiliser use at macro level. This is primarily because micro level findings identify irrigation and cropping pattern as the main variables influencing fertiliser use among different crops and farms. The same variables are used in macro models. In the process, the importance of production-response differences and untapped potential for fertiliser use was ignored. Similarly, the investigation of response functions at micro level was dominated by issues like what ought to be done rather than trying to understand why cultivators used fertilisers the way they had used. Apart from these limitations, the studies also suffer due to the fact that they were carried out when fertiliser use was low, price environment was favourable to farmers and a sizeable proportion of potential for fertiliser use was untapped. Since then not only fertiliser use has grown substantially but the environment in which it was growing has also undergone considerable changes. Thus, the need to re-examine the factors governing fertiliser use—especially its adoption, diffusion and continuous use—in areas like western tribal belt of Madhya Pradesh, which are still outside the purview of growth becomes vital.

Two different sets of data were used in the study. First, discussions with functionaries of two important institutions engaged in dissemination of farm technology (the state extension network) and supply of fertilisers in Madhya Pradesh (the cooperative supply structure). Two districts from western tribal belt of Madhya Pradesh (Dhar and Badwani) were selected for understanding the working of these institutions. Efforts were made to understand the linkages between the district level institutions and their apex organisations with respect to (a) constraints in transfer of agricultural technology and (b) the smooth supply of fertilisers. As the major concern of the study is to understand the demand-generating forces influencing fertiliser use in the western tribal belt, the second set of data was relating to farmers' fertiliser use patterns. For the purpose, two predominant tribal districts, Badwani and Dhar, were selected for micro-level study. For a proper coverage, two blocks were selected from each district. The selection of blocks was based on high concentration of tribal population and the levels of fertiliser use. In Dhar district, Kuksei represents high fertiliser use and high tribal population taluka while Dahi represented low fertiliser using high tribal population taluka. In Badwani district, Sendhwa represented high fertiliser using and Pati, a low fertiliser using taluka. From each selected block, three villages were selected, one each on the basis of availability of low vis-à-vis high levels of irrigation and predominant crops, foodgrains vis-à-vis cotton, soyabean or groundnut. In all, six villages were selected from each district. 301 randomly selected farmers from these villages were the units of observation.

Fertiliser Use Environment

Bhil, Barela, Patelia, Bhilala and non-tribal Patidar are the main inhabitants of the region. A large part of tribal-dominated area is unirrigated, highly undulating, and rocky with thin and poor quality of top soil. On the other hand, Nimar valley, inhabited by Patidars, is highly fertile and has perennial irrigation facility. The study area also falls under low rainfall region and monsoon failure is a quite frequent phenomena here. Soils in the area are capable to support cultivation only during kharif season. This tribal belt is characterised by extensive nitrogen deficiency while the levels of phosphate and potash are high in the soil. Consumption of nitrogen, therefore, is an agronomic necessity. This fact significantly influences the fertiliser use that is dominated by nitrogenous fertilisers. Although the per hectare consumption in the study districts—Dhar and Badwani at about 46 kg—is higher than the state average, the high consumption is concentrated in irrigated Nimar valley that is inhabited by non-tribal Patidars. In remote tribal areas like our study villages, the consumption is about a fourth of this.

Unlike irrigated Madhya Pradesh, supply push by the fertiliser distribution system is the main source of growth in fertiliser use in this area. Intensity of irrigation here is low and spread of HYVs is extremely poor. In fact, the available seed-fertiliser technology has bypassed the remote tribal areas. The change in the form of technologically-driven agriculture, though restricted to a minuscule of area, is not entirely an outcome of state intervention either. It happened because of seasonal migration. Seasonal migration for tribals is not coping with scarcity alone it is experiencing development that has bypassed them. What a migrating tribal learns during his stay as a labourer either on irrigated farms of Nimar plains or on the farms of irrigated HB cotton growers of Khandesh or on the fields of irrigated paddy and sugarcane growers in south Gujarat is the main source of agricultural technology transfer.

IFFCO is the lead fertiliser supplier for the region. The apex-distributing agency, the State Cooperative Marketing Federation, lifts allocated fertiliser from IFFCO to distribute it to farmers through their distribution channels. Although two different marketing channels exist in the region but it is the cooperative distribution channel that is the major distributor of fertilisers (85%) and the share of private channels is marginal (15%). This is mainly because sale of urea was under monopoly control of cooperatives till early 2004. There are over 440 sales-points of fertilisers in Dhar and Badwani districts, about 60 per cent of these are cooperative points that are located in the villages. The private sales-points selling fertilisers other than urea are located at taluka headquarters. The strengths of cooperative channel are (i) inexpensive marketing credit provided to village cooperatives by district cooperative banks, (ii) apex marketing agency's attention to placement logistics rather than reducing inventory carrying cost, (iii) coordination between different agencies, especially between fertiliser producers, village society and district Marketing Federation, and (iv) advance lifting and storage by apex marketing agency that works as a supply push to increase the sales.

In absence of a substantial growth in irrigation in this area and limited scope of diffusion of HYVs, the supply side has become an important force in generating growth in fertiliser use. An important strength of the cooperative distribution channel is its presence at the village level. Facilities of cheap marketing credit to cooperatives and production credit to farmers for purchase of fertilisers through cooperative societies is aggressive marketing strategy in the cooperative supply line. Credit limits fixed by district cooperative banks help movement of fertilisers from apex to village sales-points in the cooperative marketing structure. But a low coverage of working credit to farmers owing to default, lack of land records and at times indifferent attitude of officials of cooperatives for creating additional demand are responsible, at present, for system's inability to vigorously push fertilisers.

Fertiliser Use by Sample Farmers

Out of 301 sample farmers, 262 have applied fertilisers during 2001–02, 268 used in 2002–03 and 270 farmers used fertilisers in 2003–04. In all, 18 farmers never used fertilisers while 7 farmers after adopting fertilisers discontinued its use for last three years. It means that out of 301 farmers, 39 did not apply fertilisers in first year, 33 and 31 farmers respectively did not apply fertilisers in second and third year. Out of total 262 farmers who had applied fertilisers in 2001–02, 255 applied fertilisers continuously all the three years. Interestingly, there are 7 farmers who adopted fertilisers as late as 2001–02. In terms of nutrient use per acre, the average NPK use in the tribal area is 23.8 kg, which is less than one-fourth of the per acre use that is taking place in non-tribal areas in the study districts. Those farmers who have had a long experience of using fertilisers are applying three times more nutrients than those farmers who had started using fertilisers recently. The application of fertilisers by different size class of farmers reveals that small farmers apply relatively more nutrients per unit of area than the other farmers. In a sense, the small farmers are early adopters and the medium size farmers are late adopters of fertilisers.

Adoption of Fertilisers

Unlike highly irrigated areas where adoption of fertiliser use has a long history, in our sample of tribal farmers, adoption of fertiliser use started as late as 1970. The important crops fertilised by early adopters were maize, groundnut, cotton and wheat. Till 1980, only a few farmers adopted fertiliser use. It was only between 1992 and 2003 that there was sizable increase in fertiliser adoption; it took over 20 years for first 50 per cent of farmers to adopt fertilisers, while in next 10 years rest 50 per cent farmers adopted fertilisers. The early adopters started fertiliser use with maize, cotton, wheat, groundnut and chilli, the late adopters started fertilising their crops with soyabean and bajra. Interestingly, bulk of early fertiliser adopters on maize were unirrigated farmers. A sizable area under irrigation in crops that were fertilised was in the case of soyabean, wheat and cotton. For rest of the crops, less than 8 per cent of the crop area under fertilisers was irrigated. When fertiliser use was adopted for the first time on groundnut, chillies and bajra, it was adopted on totally unirrigated area. Urea was the main fertiliser used by the farmers when fertilisers were first adopted, followed by single super phosphate and diammonium phosphate. None of the farmers used potash when applied fertilisers for the first time. In terms of nutrient use per unit of area, highest use by first time adopters was on cotton, 34 kg per fertilised acre. The intensity of fertiliser use on the crop on which maximum number of farmers adopted fertilisers, maize was only 14.6 kg per acre. Surprisingly, fertiliser use on two other irrigated crops soyabean and wheat by first time adopters was only 16 kg per acre and 13.7 kg per acre respectively. In fact, except for cotton, there is no significant variation in fertiliser use per unit of area for first time users. This indicates that the farmers used fertilisers cautiously when they fertilised their crops for the first time; use was confined only to one crop and at reasonably low rates.

The tribal farmers were well aware of the nutrient deficiency but the physical condition of their land—highly undulating land with no irrigation facilities and thin top soil—turned them into late adopters of fertilisers. The experiences of first-time adopters of fertiliser use have been quite encouraging but not without dissent. Increased production due to fertiliser use, improvement in quality of production and crop maturing early are the most frequently reported experiences given by substantial number of the farmers. A minuscule of farmers did report negative experiences like poor crop response, rapid loss of moisture owing to fertiliser use and harmful effects on crop. But, positive effects of fertiliser use dominate the tribal psyche.

Diffusion of Fertilisers

After adopting fertiliser use on maize, 163 farmers diffused it to 299 other crops giving average less than 2 crops fertilised by them over time. In the case of 5 growers who adopted fertilisers for the first time on groundnut, diffused fertiliser use to another 12 crops giving an average additional 2.5 crops fertilised over time. This means that for some crops like maize a large number of farmers adopted fertilisers with small time lag but relatively slow rate of diffusion. For groundnut and wheat, though adoption of fertilisers was relatively late, the diffusion of fertiliser use to other crops was relatively faster. This contrast in diffusion process emerges clearly when we observe that less than 16 per cent of growers who adopted fertilisers on maize, cotton or jowar, did not diffuse fertilisers to pulses and bajra at all. On the other hand, a very large proportion of growers diffused fertiliser use on maize, cotton and wheat when they adopted fertilisers with bajra.

Observing the adoption and diffusion of fertiliser use on three important crops maize, jowar and cotton, it is noteworthy that while early adopters fertilised mainly maize, jowar, cotton and wheat, and adding over the years soyabean, bajra, gram and sometimes kharif pulses, the late adopters mainly started fertiliser use with maize, cotton, soyabean, chillies, wheat and kharif pulses. It is only after 1989 that urea use significantly increased, otherwise between 1970 and 1989, there was no significant difference in diffusion of urea, DAP and SSP. From an average 1.5 crop fertilised till 1989, the average number of crops fertilised by farmers has increased to 3. Nevertheless, there is high nutrient imbalance with no use of potash by any of the sample farmers. The phosphate and nitrogen ratio even in 2003 is 1:3 rather than 1:2. Over the period of time, there is significant variation in diffusion process across crops. While diffusion of fertiliser use on maize (72%), cotton (54%), jowar (41%) and wheat (34%) is significant, in the crops like kharif pulses, gram and bajra, less than 5 per cent farmers have diffused fertiliser use. The diffusion of fertiliser use on soyabean, groundnut and chillies is also low. If we assume that first-time adopters kept on fertilising their the crops over the time at the same level, the total NPK use due to diffusion process could be over 22 thousand kilogram in the sample. This is significantly lower than the actual use of 28 thousand kg in 2003–04. This means that overtime rates of fertiliser application increased significantly.

Factors Influencing Adoption Process

None of the social processes including tenancy is influencing the fertiliser adoption behaviour but economic indicators like better soil condition, higher proportion of plain agricultural land, non-migrant, relatively educated head of the household, large farmers, farmers with more irrigation, rich farmers and farmers whose proportion of agricultural income in total income is higher are relatively early fertiliser adopters. These findings are relevant in many ways. First, they indicate that farm condition rather than farmer's condition influences fertiliser adoption behaviour. Second, those engaged in livelihood struggles that distract them from their agriculture like relatively more labouring around the village and seasonal migration, are poor agriculturists on whose farm technology transfer in the form of fertiliser use is constrained by lack of resources. Third, better farm conditions do positively influence adoption of fertiliser use and hence avenues for agricultural technology transfer. Lastly, diffusion of fertiliser use to other crops after its adoption is also significantly influenced by the economic processes rather than by social conditions.

Sources of Growth in Fertiliser Use

Fertiliser use of the sample farmers has grown from 24.8 thousand kg in 2001–02 to 28.4 thousand kg in 2002–03. The use thereafter stagnated at 28.1 thousand kg in 2003–04. The sluggishness in use in 2003–04 has been an outcome of fall in use in all the crops, except chillies and soyabean. Despite the sluggish growth and fall in NPK use, three crops—cotton, wheat and jowar—account for two-thirds of total use of NPK. The increase in NPK use between 2001–02 and 2003–04 on three crops—cotton, chillies and wheat—accounts for about two-thirds of growth in fertiliser use. Groundnut and kharif pulses had shown a decline in fertiliser consumption while growth in fertiliser use on gram was nil. The contribution of different sources of growth in fertiliser use has been analysed for crops that have shown substantive growth, moderate growth and negative growth. Following points emerge from the analysis:

Growth pattern reveals that in wheat, chillies and cotton total increase in NPK use was substantial; in maize, jowar, soyabean and bajra the growth in NPK use was moderate while in kharif pulse, groundnut and gram the growth in NPK use was negative or stagnant.

Farmers continuously applying fertilisers during all the three years contribute to total growth in NPK (a) substantially in the case of chillies and pulses, (b) moderately in the case of maize and jowar, (c) low in the case of groundnut, and (d) negative in the case of bajra and soyabean.

A substantial reduction in growth in fertiliser use has taken place owing to discontinuation of use in the case of maize and bajra, during 2003–04 and in the case jowar, pulse and groundnut for both 2002–03 and 2003–04.

The sources of growth vary significantly across crops. (a) For wheat it is new adopters that created substantial growth in NPK use while in the case of cotton and chillies it is continuous users as well as new adopters that generated growth in NPK use; (b) in the case of maize, jowar, soyabean and bajra, the new adopters did substantially contribute to growth in NPK use; and (c) in the case of low growth crops kharif pulses, groundnut and gram, low contribution of continuous users as well as cancelling out of effect of dropout and new adopters resulted in low growth in NPK use.

Untapped Potential for Fertiliser Use

A large untapped potential in wheat and cotton, has, thus, been owing to discontinuation of use. In the moderate growth group, maize, jowar, soyabean and bajra, the untapped potential is significantly high owing not only to discontinuous use but also owing to low use by continuous users. The negative growth crops like pulses, groundnut and gram have demonstrated not only low growth in the use by continuous users but also a high discontinuation of use. In this group of crops, not only adoption is low and the continuous use is contributing marginally but also the discontinuous use is very high.

The analysis also reveals that about 32.4 thousand kg of additional NPK use can be expected in near future if extension system tries to minimise the inefficiency in fertiliser use. About 80 per cent of total potential is remaining untapped owing to inefficient use by continuous users, 16 per cent remains untapped owing to non-users and 4 per cent owing to infrequent users who dropout of fertiliser use frequently. This expected addition to fertiliser use is about 49 per cent of the total potential for NPK use of the sample farmers. Thus, in western tribal belt of Madhya Pradesh, non-adoption of fertilisers, non-use of fertilisers on a crop and infrequent fertiliser use is not a serious problem. Bulk of the inefficiency is owing to fertiliser use at relatively low rates.

The extension agency need not become highly diffused in the process because:

> If non-users of jowar, kharif pulses and soyabean start using fertiliser on these crops, the additional growth in fertiliser use shall be over 70 per cent of growth in fertiliser use that could be achieved if all the non-users are encouraged to start using fertiliser. It implies that extension system can become selective in propagating fertiliser. They could start with jowar, kharif pulses and soyabean and over time move to other crops.
>
> If infrequent users of jowar, maize and cotton are encouraged to use fertilisers regularly, the additional growth by these farmers shall be over 60 per cent of total growth that could be created due to regular use by all the infrequent users.
>
> If continuous users of maize, cotton, jowar and wheat increase their rates of application, the growth of NPK use would be about 90 per cent of total growth due to improvement in rates on all crops.
>
> This also implies that improving the efficiency of fertiliser use of continuous users on cotton, maize, jowar and wheat is key to achieving over 80 per cent of untapped potential. The task of the extension system is now much easier.
>
> A portion of the inefficiency is on account of farmers' perception that optimum use on a crop is actually lower than what is recommended by the extension agency, even under very difficult conditions of remote tribal areas. But a part of the inefficiency could also be owing to variations of the conditions that influence fertiliser use; the farm conditions.

Inefficiencies amongst Infrequent Users

While examining the issue as to why on some farms fertiliser use was continuous, on some it was infrequent and on some others fertiliser use has yet to start, it is found that among social factors, none other than education is associated with the pattern of fertiliser use. It means that whether one is a tribal or Scheduled Caste or Patidar, there is no significant variation in the pattern of fertiliser use amongst the three; size of family and the family structure are also not associated with the patterns of fertiliser use. But among literates, a significantly larger proportion of farmers are continuous users rather than discontinuous users, whereas among illiterate, significantly· larger proportion of households are non-users and infrequent users.

Among economic variable fragmentation of land holdings and tenancy has no association with pattern of fertiliser use. On the other hand, longer experience of fertiliser use, larger size of holdings, better soil condition, farms with large proportion of area plain, high income households, households where agriculture is substantially contributing to family income and farms with larger proportion of area under irrigation have relatively more continuous fertiliser users. While those with few years of experience of fertiliser use, small size of holdings, rocky soils with larger proportion of area undulating, low income households with larger proportion of income originating from labour and migration, and farms with no irrigation facilities have relatively more non-users and infrequent users of fertilisers. More importantly, two support factors: fertiliser use technology and input-output market support are significantly and positively associated with the pattern of fertiliser use. But credit support and extension support are not able to influence significantly the pattern of fertiliser use. This, in a sense, reflects on poor access to institutional finance and poor quality of extension support in the area. All those who know more about the sources of fertiliser nutrients and the time and method of use of these nutrients as well as better access to institutional marketing are relatively more continuous users.

What the above findings clearly reveal is, like adoption of fertilisers, its patterns of use too are governed by farm characteristics rather than farmer's characteristics. Improving (a) the soil moisture of farm, (b) ways to apply fertilisers, and (c) access to institutional markets for input-output, are far more important in converting non-adopters of fertilisers to adopters and changing a discontinuous use pattern towards a continuous use pattern. Accumulated experience is just not the number of years since one is using fertilisers. It also reflects on changes in attitude towards fertilisers, knowing the nuances associated with its use and experiencing productivity gains due to fertiliser use. The need, therefore, is to calibrate the extension messages, access to credit on the one hand, and productivity gains through public investment on the other, for improving the growth in fertiliser use.

Inefficiencies amongst Continuous Users

Interestingly, over a third of the respondents believe that soil moisture constraints do not allow them to increase their fertiliser use. But over half of the farmers either show lack of extension support in the form of know-how of fertiliser use or there is utter confusion relating to wrong perceptions about fertiliser use. The misconceptions range from quality and fertility of soil deteriorate with fertiliser use to one needs to increase fertiliser use every year to pest attacks increase after fertiliser use. Consequently, these farmers prefer to use fertilisers at relatively low rates. The analysis relating to effect of price increase of fertiliser on its use indicates that farmers' short-term reaction to price rise by a third of farmers to reduce the rate of application was rational but their long-term reaction that they would continue low fertiliser use despite a net loss due to yield reduction for a subset was irrational.

What this clearly brings out the need for a rational extension message, suited for the tribal environment. It also brings out clearly the need for larger on-farm experiments to demonstrate the optimality of fertiliser use. This tribal area represents relatively poor endowment and still poor agricultural infrastructure. All the major crops like maize, jowar, pulses are grown largely under unirrigated conditions. Some irrigation that exists is applied for long duration crops like cotton or rabi cultivation, mainly wheat. The level of fertiliser use is generally lower by about 60 per cent of the recommended rates for different crops (except for pulses and groundnut). Despite the fact that fertiliser prices for these farmers is a non-issue, the fertiliser marketing environment—whereby urea monopoly sale through cooperatives has broken—has set in at a time when the farmers economy has faced serious problem due to recurring monsoon failure. As a result, large number of farmers have chosen to raise price issue to voice their grievances against the neglect of the area, especially with respect to agricultural infrastructure level of available technology, low productivity and high cost and high risk associated with private investment in irrigation. Nonetheless, independent of fertiliser prices, a number of other and equally important issues are thrown up by our investigation. Forces like lack of irrigation facilities and consecutive crop failure, poor soil with low agricultural productivity have resulted in creating perceptions that low rates of fertiliser use is a better risk management device in uncertain yield-response environment.

Conclusion and Policy Implications

The base that generated the past growth in fertiliser use has been narrow in the remote tribal belt of western Madhya Pradesh. This growth has happened mainly due to improvement in the rates of fertiliser application on the farms of continuous users. The growth could have been faster but for farmers who were not fertilising all of the major crops despite having adopted fertilisers, infrequent use of fertilisers and continuous users fertilising their crops at sub-optimal rates.

The growth in fertiliser use by continuous users was maintained at low rate because of the fact that effective fertiliser rates are still about a half of the optimum rates for major fertilised crops cotton, maize, wheat and jowar. It is a typical case of low use slowly moving towards its potential; the rightward shift in fertiliser demand function was missing owing to the fact that there were no upward shifts in production functions. This growth has still not attained its vigour, as a significant proportion of untapped potential still exist on major fertilised crops. On the other hand, any shift in cropping pattern in favour of these high fertiliser-consuming crops cannot be expected without quantitative improvement in irrigation, diffusion of fertiliser responsive varieties and avenues of assured markets for output. As high risk associated with private investment in irrigation has constrained the diffusion of HYVs in this area, Public investment in (a) soil-water conservation, and (b) adopting HYVs suited for the region is not coming forth. This, if happens, will increase the existing potential of fertiliser use by upward shift of production functions.

Sah (1982) has argued that adoption and diffusion of fertiliser use is governed by size and certainty of returns on fertiliser. Irrigation creates an agro-economic environment in which not only size of returns due to fertiliser use increases but also becomes certain. This attracts farmers not only to diffuse fertilisers on unfertilised area faster but also do so with relatively higher rates. This is mainly because additional irrigation improves the potential for fertiliser use. Irrigation also brings cropping pattern changes whereby the low fertiliser using pulses and bajra are replaced by fertiliser responsive crops like soyabean, cotton and maize. Additional irrigation also increases the cropping intensity by reducing rabi fallow by fertiliser responsive wheat and late groundnut. But adding irrigation is only one of the prerequisites for a rapid growth in fertiliser use, equally important is to create a pool of fertiliser responsive seed varieties suited for the location. Such investment is necessary for both irrigated as well as unirrigated crops.

Contours of future growth in fertiliser use in tribal areas are, thus, quite clear. Two processes need to be simultaneously initiated. First, creating additional potential for fertiliser use, and second, minimising the inefficiencies in fertiliser use of existing users. A large part of inefficiency in the patterns of fertiliser use is owing to perceptions of farmers that the returns to added fertilisers are not only low but also highly varying. Since there is a hierarchy of returns and their certainty across crops, some crops are fertilised regularly, some infrequently and some never. These perceptions encourage fertiliser adopters not to fertilise all the crops they grow. They also influence some farmers to discontinue its use on some crops. On the other hand, these perceptions are also responsible for continuous users to fertilise their crops at lower than optimum rates. Following processes are inevitable to addressing these inefficiencies.

First, perceptions of farmers about size and certainty of returns can be improved by demonstrating them that through proper farm practices, the returns can be improved as well as risk associated with fertiliser use could be minimised. These demonstrations need to be organised on a large-scale both for irrigated as well as unirrigated crops.

Second, improving farmers' fertiliser use technology by improving their knowledge is an important input for minimising inefficiencies. Time and method of fertiliser use, balanced versus unbalanced use, impact of micro-nutrients on the response of chemical fertilisers, synergetic effect of farm yard manure and chemical fertilisers, and fertiliser use at optimum rates are components that could change significantly the economics of fertiliser use. Revitalising the soil testing services that have almost lost their bearing on the region (see chapter 7) is one way to improve farmers' fertiliser use technology.

Third, in a spatial dimension where rainfed agriculture has remained outside the mainstream of growth in fertiliser use, the fertiliser distribution system and the research-extension system tends to cater to the needs of high potential areas only. As a result, the presence of fertiliser distribution and research-extension systems is symbolic; merely to facilitate the low demand. Creating additional demand has remained outside the purview of these two agencies. In the process, neither the distribution network considers expanding their business economically viable nor the agronomic complexities of these difficult areas receive adequate importance in the priorities of research and extension network. The importance of extensive adaptive research cannot be overemphasised in order to identify sources of response gaps between research station and cultivators' fields. This is especially a prerequisite for crops where fertiliser use is not common. This needs to be followed by a meaningful promotional campaign. These tasks in the tribal region have been solely undertaken by the state. Fertiliser industry can play an important role in this respect. After the decontrol of fertiliser distribution in this region in 2004, the urea marketing has become commercial. This commercialisation breeds, as experiences of other areas show, concentration of use in traditionally high fertiliser consuming areas. Since, the traditional areas cannot sustain a rapid growth year after year, pressures on industry to sell fertilisers in non-traditional areas have to be exerted. An instrument for such pressure on industry could be *supply push* that could calibrate industry's efforts to create additional demand and expand the distribution network in non-traditional fertiliser consuming areas.

Fourth, basic investment in agriculture in remote tribal areas, in the form of soil-water conservation, is amongst the most desirable involvement by state. Adequate investment in irrigation, drainage and gully plugging, soil conservation helps in two ways. It insulates agriculture from vagaries of monsoon and also opens avenues for an extensive technological transformation in tribal agriculture. If happened efficiently, it results in production function shifting upwards and input demand function shifting to the right. In other words, it creates additional potential for fertiliser use. But such public investments are riddled with serious problems like lack of people's participation, poor maintenance and iniquitous distribution of gains from such investment (Sah and Bhatt, 2004). The need, therefore, is to initiate processes that result in transparency and accountability in such investment by encouraging community participation in such ventures. It has also been argued that inclusion of civil society in these ventures is one way of making them economically viable and community friendly.

Fifth, the uncertainty in fertiliser price environment has distorted both micro- as well as macro-processes that have made some fertilisers dearer (phosphate and potash) than others (nitrogen). In spite of this, fertiliser use in early 2000s in tribal regions has grown substantially. The major reason for this growth has been low rates, much lower than the optimal for the region. Consequently, even in the changed price environment, new optimum is much higher than the effective rates of cultivators. These micro findings indicate that in low fertiliser using areas only an insignificant portion of year-to-year variations in fertiliser use towards its potential are explained by the real price of fertilisers. This means that in tribal region where both spread of fertilisers on different crops is poor and rates of fertiliser application are low cannot be converted to growth-generating sources through manipulations of prices in favour of cultivators. The non-price factors are far more important.

Lastly, nutrient application at much lower than the recommended rates implies that fertilisers are being applied with serious misconception about nutrient dynamics and its balanced use. The failure of the extension systems to appreciate farmers' rationale for their fertilisers use behaviour, results not only in lack of demand for extension services but also in lack of faith on these services. Understanding how perceptions are formed as well as the constraints the farmers of the region face is prerequisite for influencing farmer's fertiliser use behaviour.

The task of meeting need-based targets of fertiliser use, thus, involves following policy interventions: (a) improving fertiliser use efficiency in the case of irrigated wheat, jowar, cotton and maize farmers; (b) developing new fertiliser responsive varieties for both rainfed and irrigated agriculture; (c) strengthening the extension and supply arrangements in the region; and (d) investment on soil and water conservation in the tribal areas.

All these are challenging but essential tasks for improving efficiency of fertiliser use. Future growth in agricultural productivity in the tribal areas cannot be sustained by only more and more yield increasing inputs. Their efficient use cannot be overemphasised.

Bibliography

Abraham, T.P., "Optional Fertiliser Dressing and Economics of Manuring", *Indian Journal of Agricultural Economics,* Vol. 20 (2), April-June 1965, pp. 1–20.

Acharya, S.S. and Jogi, R.L., "Farm Input Subsidies in Indian Agriculture", IDSJ Working Paper No. 140, Institute of Development Studies, Jaipur, 2004.

Babaria, C.H., "Fertiliser Use and Crop Response Studies: A Review", *Arth Vikas,* Vol. 14 (1), January-June 1978, pp. 53–62.

Bains, S.N., Prashad, Rajendra and Bhatia, P.C., "Use of Indigenous Material to Enhance the Efficiency of Fertiliser Nitrogen for Rice", *Fertiliser News,* Vol. 19 (3), March 1971, pp. 30–52.

Bal, H.K. and Bal, H.S., "Fertiliser Demand for Wheat Crop in Punjab", *Indian Journal of Agricultural Economics,* Vol. 23 (4), October-December 1973, pp. 68–76.

Bank of India, "Fertiliser Consumption in India: Problems and Prospects", *Bank of India Bulletin,* Vol. 11 (3 and 4), March-April 1973.

Barker, R. and Hayami, Y. "Price Support Versus Input Subsidy for Food Self-Sufficiency in Developing Countries", *American Journal of Agricultural Economics,* 58 (4), 1976.

Barker, R., Bennagen, E. and Hayami, Y. "New Rice Technology and Policy Alternatives for Food Self-Sufficiency" in eds., International Rice Research Institute, *Economic Consequences of the New Rice Technology,* Manila, Philippines, 1978, pp. 337–61.

Bliss and Stren, N.H., *Palanpur: Economy of Indian Villages*, Oxford University Press, Madras, 1982.

Brown, D.D., *Agricultural Development in India's Districts*, Harvard University Press, Cambridge, 1971.

Chaudhary, B.N. et al., "Problems of Farmers in Relation to Fertiliser Use: An Investigation in Delhi Villages", *Fertiliser News*, Vol. 17 (12), December 1972, pp. 51–56.

Chaudhary, K. and Maharaja, M., "Acceptance of Improved Practices and their Diffusion among Wheat Growers in Pali District of Rajasthan", *Indian Journal of Agricultural Economics*, Vol. 21 (1), January-March 1966, pp. 199–208.

Chaudhri, A.K. and Sirohi, A.S., "Allocation of Fertilisers Among Crops and Regions in Uttar Pradesh", *Indian Journal of Agricultural Economics*, Vol. 28 (3), July-September 1973, pp. 46–61.

Cummings, R.W. Jr. et al., "New Agricultural Strategy Revisited", *Economic and Political Weekly*, Vol. 3 (43), October 26, 1968, pp. A15–23.

Cummings R.W. and Ray, S.K., "1968–69 Foodgrain Production: Relative Contribution of Weather and Technology", *Economic and Political Weekly*, Review of Agriculture, September 1965.

Dahiya, R.C., "Prediction of Fertiliser Consumption in India", *Indian Journal of Agricultural Economics*, Vol. 22 (1), January-March 1967, pp. 108–110.

Dantwala, M.L., "Agricultural Policy in India since Independence", in C.H. Shah and C.N. Vakil (eds.) *Agricultural Development of India: Policy and Problems*, Orient Longman, 1977, pp. 3–29.

——, "From Stagnation to Growth", *Indian Economic Journal*, Vol. 18 (2), October-December 1970, pp. 165–88.

David, C.C. and Barke, R., "Modern Rice Varieties and Fertiliser Consumption" in eds., International Rice Research Institute, Economic Consequences of the New Rice Technology, IRRI, Manila, Philippines, 1978, pp. 175–211.

Desai, D.K. and Sharma, B.M., "Technological Change and Rate of Diffusion", *Indian Journal of Agricultural Economics*, Vol. 21 (1), January–March, 1966, pp. 141–154.

Desai, D.K., "Intensive Agricultural District Programme: Analysis of Results" in Pramit Choudhuri (ed.), *Readings in Indian Agricultural Development*, George Allen &. Unwin Ltd., London, 1972, pp. 140–154.

Desai, D.K. and Doshi, S.P., "Economics of Fertiliser Use", *Indian Journal of Agricultural Economics,* Vol. 17 (2), April-June 1962, pp. 65–73.

Desai, G.M. and Singh, Gurdev, *Growth in Fertiliser Use in District of India: Performance and Policy Implications,* CMA, Indian Institute of Management, Ahmedabad, 1973.

Desai, G.M. and Mellor, J.W., "Changing Basis of Demand for Fertilisers", *Economic and Political Weekly,* Vol. 4 (39), September 27, 1969, pp. A175–87.

Desai, G.M., "A Critical Review of Fertiliser Consumption after 1974–75 and Prospects of Future Growth", *Fertiliser News,* Vol. 23 (7), July 1978, pp. 7–13.

——, "Comment on Modern Rice Varieties and Fertiliser Consumption" in eds., International Rice Research Institute, Economic Consequences of the New Rice Technology, IRRI, Manila, Philippines, 1978, pp. 213–20.

——, "Fertiliser in Indian Agricultural Development" in C.H. Sah and C.N. Vakil (eds.) *Agricultural Development of India: Policy and Problems,* Orient Longman, Bombay, 1979, pp. 377–426.

——, "Measuring Promotional Effectiveness", *Fertiliser News,* Vol. 20 (4), April, 1975, pp. 3–8.

——, "Some Observations on Economics of Cultivating HYVs of Rice in India", *Arth Vikas,* Vol. 7 (2), July 1971, pp. 1–8.

——, "Fertiliser Policy: A Perspective on Price and Subsidy Issue", in eds., Vidya Sagar, *Fertiliser Prices: Issues Related to Subsidy,* Classic Publishing House, Jaipur, 1993.

——, "Rapporteur's Report: Impact of Increase in Input Prices on Profitability and Production", *Indian Journal of Agricultural Economics,* Vol. 31 (3), July-September 1976, p. 138–56.

——, Growth of Fertiliser Use in Indian Agriculture: Past Trends and Future Demand, Occasional Paper No. 24, Department of Agricultural Economics, Cornell University, Ithaca, New York, July 1969.

——, Nitrogen Use and Foodgrains Production, India 1973–74, 1978–79 and 1983–84, Department of Agricultural Economics, Cornell University, Ithaca, New York, 1973.

——, "Reporter's Report: Agricultural Input Supply Systems Including Marketing", *Indian Journal of Agricultural Economics,* Vol. 29 (4), October-December 1973, pp. 158–72.

——, Sustaining Rapid Growth in India's Fertiliser Consumption: A Perspective Based on Use, International Food Policy Research Institute, Washington, USA, August 1982.

——, Understanding the Process of Growth in Fertiliser Consumption: A Conceptualisation, International Food Policy Research Institute, Washington, USA, 1996.

——, Policies for Growth in Fertilizer Consumption, The Next Stage, Working Paper No. 602, Indian Institute of Management, Ahmedabad, 1986.

——, "Issues and Themes in Growth of Fertiliser Use in India", *Journal of Indian Society of Agricultural Statistics,* Vol. 43 (1), April 1991.

Desai, G.M., Chary, P.N. and Bandyopadhyay, S.C., *Dynamics of Growth in Fertiliser Use at Micro Level,* CMA, Indian Institute of Management, 1979.

Desai, G.M. and Stone, B., "Fertiliser Market Development and National Policy in China and India", (mimeo), Indian Institute of Management, Ahmedabad, 1987.

Dhawan, B.D., "'Impact of Irrigation on Agricultural Output: A Case Study of Tamil Nadu", *Economic and Political Weekly,* September 1983.

——, "Indian Irrigation: An Assessment", *Economic and Political Weekly,* May 7, 1988.

——, *Growth of Productivity in Indian Agriculture*, Cornell University, New York, 1976.

Dhillon, B.S. and Sankhayan, P.L., "An Analysis of Impact of Factor Product Price and Credit Availability on Demand for Fertilisers in Punjab", *Indian Journal of Agricultural Economics*, Vol. 32 (4), October-December 1977, pp. 99–107.

Directorate of Economics and Statistics, Estimates of Area and Production of Principal Crops in India, An Annual Publication, GoI, New Delhi.

——, Farm Harvest Prices of Principal Crops in India, GoI, Ministry of Agriculture and Irrigation, New Delhi, 1977.

——, Intensive Agricultural District Programme, 1968–69 to 1970–71, Vol. I, Fifth Report, GoI, New Delhi, 1975.

Dogra, Lata, "Economics of Fertiliser Use", *Economic and Political Weekly,* Vol. 4 (37), September 1969, pp. 1485–87.

Fertiliser Association of India, Fertiliser Statistics, Annual Publication of Fertiliser Association of India, New Delhi, Various Issues.

——, "Consumption Production, Imports and Distribution of Fertiliser in India, 1975–76 to 1978–79", *Fertiliser News,* Vol. 21 (1), January 1976, pp. 36–39.

——, "Fertiliser Consumption and Availability in India, 1976–77 to 1983–84", *Fertiliser News,* Vol. 21 (8), August 1976, pp. 3–6.

——, "Fertiliser Urea: Its Efficient Use" *Fertiliser News,* Vol. 22 (9), September 1977, pp. 3–18.

——, "Impact of Fertiliser Price on Consumption of Fertilisers", *Fertiliser News,* Vol. 22 (11), November 1977, p. 17 and 27.

——, "Village Adoptive Research Cum Demonstration Programme on Integrated Nutrient Supply System", *Fertiliser News,* Vol. 22 (3), March 1977, pp. 4–6.

——, A Study on Fertiliser Demand and Marketing, FAI, New Delhi, 1974.

——, Economics of Fertiliser Use. Proceedings of the Group Discussion, May 30–31, 1978, FAI, New Delhi, July 1975.

——, FAI-FAO Seminar on Optimising Agricultural Production Under Limited Availability of Fertilisers, 1974. Seminar Proceedings, FAI, New Delhi, 1975.

——, Trends in Consumption and Production, FAI-IFDC Seminar on Trends in Consumption and Production of Fertilisers, December 1–3, 1977, New Delhi.

Food and Agricultural Organization, Statistics of Crop Response to Fertilisers, FAO, Rome, 1966.

Girdharilal, "Factors Affecting Economics of Fertiliser Use", *Fertiliser News,* Vol. 14 (2), February 1969, pp. 21–25.

——, "Practical Aspects of Economic of Fertiliser Use", *Fertiliser News,* Vol. 19 (3), March 1974, pp. 25–29.

Goel, S.K., "Selection of Profitable Doses of Nitrogen for Rice", *Fertiliser News,* Vol. 19 (2), February 1974, pp. 28–32.

Gopalakrishnan, M.D., "Productivity and Profitability of ADT-27 in Thanjavur District", *Indian Journal of Agricultural Economics,* Vol. 23 (4), October-December 1968, pp. 65–69.

Goswami, N.N. et al., "Fertiliser Recommendations on Farmer's Fields: Soil Fertility Cum Soil Survey Approach", *Fertiliser News,* Vol. 18 (3), March 1973, pp. 32–38.

Griliche, Zvi, "The Demand for Fertilisers: An Economic Interpretation of a Technical Change", *Journal of Farm Economics,* Vol. 40 (3), August 1956, pp. 591–601.

Gulati, A., "Input Subsidy in Indian Agriculture", *Economic and Political Weekly,* June 24, 1989.

Gulati, A. and Narayanan, S., "Fertiliser Subsidy', in eds., *The Subsidy Syndrome in India,* Oxford University Press, New Delhi, 2003.

Gulati, A. and Sharma, A., "Subsidy Syndrome in Indian Agriculture", *Economic and Political Weekly,* September 30, 1995.

Hayami, Y., "Demand for Fertiliser in the Course of Japan's Agricultural Development", *Journal of Farm Economics,* Vol. 45 (4), November 1964, pp. 766–80.

Hayami, Y. et al., "Price Incentive Versus Irrigation Investment to Achieve Food Self-sufficiency", *American Journal of Agricultural Economics,* 1977, Vol. 59, pp. 717–21.

Herdt, R.W. and Mellor, J.W., "The Contrasting Response of Rice to Nitrogen: India and the US.", *Journal of Farm Economics,* Vol. 46 (1), February 1964, pp. 150–61.

Herdt, R.W., "Nitrogen Response of Rice: 1968 AICRIP Trials", *Economic and Political Weekly,* Vol. 6 (13), March 27, 1971, pp. A33–A36.

Heredia, F.J., "Fertiliser Consumption: Can it be Doubled in the Next Five Years", *Chemical Age of India,* Vol. 29 (12A), December 1978, pp. Mktg. 1–5.

Hopper, W.D., "Economics of Fertiliser Use: A Case Study of Production Economics", *Indian Journal of Agricultural Economics,* Vol. 17 (4), October-December 1962, pp. 12–72.

——, "Planning Yardsticks for Fertiliser and Irrigation", *Agricultural Situation in India,* Vol. 20 (6), September 1965, pp. 463–77.

Hsu, R.C., "The Demand for Fertiliser in a Developing Country; The Case of Taiwan", *Economic Development and Cultural Change,* Vol. 21, January 1972, pp. 299–309.

Institute of Agricultural Research Statistics, India, Sample Survey of Assessment of HYVs Programme, Annual Reports for Years 1970–71 to 1975–76, Vol. II, IARS, New Delhi.

Ishikawa S., *Economic Development in Asian Perspective*, Kinrkuniya, Tokyo, 1967.

Iyer, V.N. and Sardana, M.G., "Economics of Fertilisation of Maize under Cultivators' Condition", *Fertiliser News*, Vol. 18 (6), June 1973, pp. 15–19.

Jayaraman, T.K., "Fertilizer Consumption in Gujarat State: An Inter District Analysis", *Fertilizer News*, 24 (11), New Delhi, 1979.

Jha, Dayanatha and Sari, Rakesh, Fertilizer Consumption and Growth in Semi-Arid Tropical India: A District Level Analysis, Inter National Crop Research Institute for Semi-Arid Tropics (ICRISAT), Hyderabad, 1980.

Jha, Dayanatha, Fertilizer Use and its Determinants: A Review with Special Reference to Semi-Arid Tropical India, ICRISAT, Hyderabad, 1980.

Kahlon, A.S., "Presidential Address: New Farm Technology: Its Implications in Agricultural Economics", *Indian Journal of Agricultural Economics*, Vol. 25 (4), October-December 1970, pp. 1–14.

Kahlon, A.S. et al., "A Comparative Analysis of Dry and Irrigated Farming in Ferozepur District, Punjab", *Indian Journal of Agricultural Economics*, October-December 1971, pp. 318–26.

Kahlon, A.S. and Sirohi, A.S., 'Fertiliser Use Under Constrained Supplies", *Fertiliser News*, Vol. 19 (12), December 1974, pp. 37–38.

Kanwar, J.S., "Coordinated Research in Soil Productivity and Water Management", *Fertiliser News*, Vol. 17 (6), June 1972, pp. 13–16.

Kanwar, J.S. et al., "Are Fertilisers Application to Jowar, Maize and Bajra Economical", *Fertiliser News*, Vol. 18 (7), July 1973, pp. 19–28.

——, "Balanced Fertiliser Use for Maximising Returns from Wheat on Cultivators' Fields", *Fertiliser News,* November 1972, Vol. 17 (11), pp. 19–30.

——, "Economics of Fertiliser Use in Rice in Farmers' Fields", *Fertiliser News*, Vol. 18 (1), January 1973, pp. 71–88.

Krishnaji, N., "Measuring Agricultural Growth", *Indian Journal of Agricultural Economics,* Vol. 35 (2), April-June 1980.

Krishnan, K.S. and Mahrotra, P.C., "Crop Response to Fertiliser Application", *Fertiliser News,* Vol. 15 (8), August 1970, pp. 23–28.

Krishnan, K.S., Soni, P.N. and Rastogi, V.S., "A Study on Irrigation and Fertilizer Responses in IADP districts", *Fertilizer News,* March 1971.

——, "Irrigation and Fertilizer Response of Wheat, Maize, Jowar and Bajra in IADP districts", *Fertilizer News,* 1972.

Leela, P., "Fertilizer Consumption in A.P.: A District-wise Analysis", *Agriculture Situation in India,* November 1985.

Mahapatra, I.C. et al., "Economics of Fertiliser Use", *Fertiliser Marketing News,* Vol. 5 (12), December 1974, pp. 1–12.

——, "Fertiliser Response of Wheat, Gram and Horse Gram Under Rainfed Conditions", *Fertiliser News,* Vol. 18 (4), April 1973, pp. 57–63.

——, "Response of Rice, Jowar, Maize, Bajra, Groundnut and Caster to Fertilisers under Rainfed Conditions on Farmers' Fields", *Fertiliser News,* Vol. 18 (8), August 1973, pp. 18–28.

Maharaja, M.H., "Demand for Fertilisers, Good Companions", Baroda, 1975.

Mellor, John W. and Lele, Uma, 'Growth Linkages of the New Foodgrain Technology', *Indian Journal of Agricultural Economics,* 29 (4), 1973.

Mellor, John W., "Increasing Agricultural Production in Early Stages of Economic Development: Relationships, Problems and Prospects", *Indian Journal of Agricultural Economics,* 17 (2), pp. 29–46, April-June 1962.

Menon, V.S., "Role of Fertilisers in Crop Production in India", *Agricultural Situation in India,* Vol. 17 (1), April 1962, pp. 27–38.

Minhas, B.S. and Srinivasan, T.N., "New Agricultural Strategy Analysed", *Yojana,* January 26, 1966, pp. 20–24.

Minhas, B.S. and Vaidyanathan, A., "Growth of Crop Output in India, 1951–54 to 1958–61: An Analysis by Component Elements", *Journal of the Indian Society of Agricultural Statistics,* Vol. 17 (2), 1965.

Ministry of Agriculture and Irrigation, Report of the National Commission on Agriculture 1976, Vol. X (Inputs), GoI, New Delhi, 1976.

Ministry of Food and Agriculture, Modernising Indian Agriculture; Fourth Report on the Intensive Agricultural District Programme (1960–68), Vol. I and II, GoI, New Delhi, 1960 and 1970.

——, Report of the Expert Committee on Manure Fertilisers, GoI, New Delhi, 1953.

——, Report of the Expert Committee on Assessment and Evaluation of Intensive Agricultural District Programme 1960–63, First Report, GoI, New Delhi, 1963.

——, Report of the Fertiliser Distribution Enquiry Committee, GoI, New Delhi, 1960.

——, Report on Committee on Fertilisers, GoI, New Delhi, 1965.

——, Report on Intensive Agricultural District Programme, 1965–66 and 1966–66, GoI, New Delhi, 1967.

——, Second Report on Intensive Agriculture District Programme, 1960–65, GoI, New Delhi, 1966.

Mishra, V.N. and Gupta, S.B.L., "The Impact of Fertiliser Use on Resource Productivity: A Study of an IADP Block", *Indian Journal of Agricultural Economics*, Vol. 52 (4), October-December 1972, pp. 311–25.

Mishra, V.N., "Fertiliser Use Efficiency in Resource Allocation: A Study of Paddy Crop in an IADP Block", *Economic and Political Weekly*, Vol. 6 (14), April 3, 1971, pp. 752–55.

——, "Farm Structure and Fertilizer Use", *Agricultural Situation in India*, Vol. 25 (10), January 1979.

Misra, V.N., "The Effect of Institutional Factors on the Use of Fertilisers", *India Cooperative Review*, Vol. 4 (4), 1967, pp. 120–27.

Mohanam, T.C., "Determinants of Fertilizer Use in Tamil Nadu: An Analysis", *Agricultural Situation in India*, September 1990.

——, "Diffusion and Intensity of Fertilizer and Manure Use in Indian Agriculture: A Statewise Analysis", *Southern Economists*, August 1987.

——, "Fertilizer Subsidy in Indian Budgets", *Industrial Economist*, Annual Number, 21 (4), June-July 1988.

——, "Growth Rates of Fertilizer Consumption: A District Wise Analysis in Tamil Nadu", *Agricultural Situation in India*, February 1989.

Mudahar, M.S., "Needed Information and Economic Analysis for Fertiliser Policy Formulation", *Indian Journal of Agricultural Economics*, Vol. 33 (3), July-September 1978, pp. 46–67.

Muralidharan, N.A. and Gangadharan, T.P., "Demand for Chemical Fertilisers in Kerala", *Fertiliser Marketing News*, Vol. 5 (9,10,11) September to November 1974.

Nagaraj, R., "Determinants of Fertilizer Use in Indian Agriculture", *Economic and Political Weekly*, Vol. 18 (3), March 26, 1983.

Nampoothiri, N.V. and Desai, G.M., "Demand versus Supply Factors in Growth of Fertiliser Use", in eds., G.M. Desai and A. Vaidyanathan, *Strategic Issues in Future Growth of Fertiliser Use in India*, ICAR, New Delhi and IFPRI, Washington DC, 1994.

Nandal, D.S. and Grover, D.K., "Input Price, Production and Profitability in Haryana", *Indian Journal of Agricultural Economics*, Vol. 31 (3), July-September 1976, pp. 86–94.

Narain, Dharm, "Growth of Productivity in Indian Agriculture", *Indian Journal of Agricultural Economics,* Vol. 22 (1), January-March 1977, pp. 1–44.

——, "Distribution of the Marketed Surplus of Agricultural Produce by Size-level of holding in India 1950–5", in eds., K.N. Raj et al., Studies on Indian Agriculture, Oxford University Press, 1988.

——, *Growth of Productivity in Indian Agriculture*, Cornell University, New York, 1986.

Narayan, Pratap, "Marketing Fertilizers Affordable for Farmers' Subsidies and Alternatives", *Fertilizer News,* October 1987.

——, "Fertilizer Pricing in India" in EL Segura et al. (ed.) *Fertilizer Producer Pricing in Developing Countries Issues and Approaches,* World Bank, Washington, January 1986.

National Committee on Science and Technology, Status Paper on Fertiliser, GoI, New Delhi, 1972.

NCAER and FAI, Fertiliser Use on Selected Crops in India, National Council of Applied Economic Research and Fertiliser Association of India, New Delhi, 1974.

NCAER, Long Term Projections of Demand for and Supply of Selected Agricultural Commodities, National Council of Applied Economic Research, New Delhi, 1962.

——, Factor Affecting Fertiliser Consumption, Problems and Policies, National Council of Applied Economic Research, New Delhi, 1964.

——, Fertiliser Demand Study: Interim Report, Vols. II to VII, National Council of Applied Economic Research, New Delhi, 1978.

——, Fertilizer Demand Survey 1975–76, National Council of Applied Economic Research, Interim Report, Vol. 1–6, New Delhi, 1978.

——, Fertilizer Use on Selected Crops in India: 1970–71, New Delhi, 1978.

——, A Study of Factors Influencing Fertilizer Use, Main Report, National Council of Applied Economic Research, New Delhi, November 1990.

Panse, V.G., "Demand for Nitrogenous Fertilisers: An Economic Study", *Indian Journal of Agricultural Economics*, Vol. 20 (3), July-September 1965, pp. 13–20.

——, "Fertiliser Trial", *Fertiliser News,* Vol. 10 (12), December 1965, pp. 19–23.

——, "Promotion and Assessment of Technological Changes in Indian Agriculture", *Indian Journal of Agricultural Economics,* January-March 1966, pp. 873–79.

——, "'Fertiliser Recommendations", *Fertiliser News*, Vol. 11 (6), June 1966, pp. 7–12.

Panse, V.G. et al., *Yardstick of Additional Production of Certain Foodgrains, Commercial Crops and Oilseed Crops,* ICAR, Institute of Agricultural Research Statistics, New Delhi, 1964.

Pant, S.P., "Impact of Fertilizer Price Increasing on Fertilizer Consumption and Agricultural Output", *Agricultural Situation in India,* Vol. 35, October 1980.

Parikh, A.K., "Complementarily between Irrigation and Fertilisers in Indian Agriculture", *Indian Journal of Agricultural Economics*, Vol. 24 (3), July-September 1969, pp. 1–18.

——, "Consumption of Nitrogenous Fertilisers: A Continuous Cross-section Study and Covariance Analysis", *Indian Economic Journal*, Vol. 14 (2), December 1966, pp. 873–79.

Parikh, Ashok and Mosely, S., 'Fertilizer Response in Haryana', *Indian Journal of Agricultural Economics,* Vol. 41 (2), April-June 1986.

Parikh, K.S. and Srinivasan, T.N., *Optimum Requirements of Fertiliser for the Fifth Plan Period,* Indian Statistical Institute and FAI, New Delhi, 1974.

Parikh, K.S. and Suryanarayana, M.H., "Food and Agricultural Subsidy", *Journal of Quantitative Economics*, Vol. 8 (1), 1992.

Parikh, K.S., "HYV and Fertilisers: Synergy or Substitution", *Economic and Political Weekly*, Vol.12 (12), March 25, 1978, pp. A 2–8.

Parikh, Ashok, "Demand for Nitrogenous Fertilizers: An Economic Study", *Indian Journal of Agricultural Economics,* Vol. 20 (3), July-September 1965.

Parthasarathy, G. and Mudaliar, M.S., "Foodgrain Prices and Economic Growth", *Indian Journal of Agricultural Economics*, Vol. 21 (2), April-June 1976, pp. 16–30.

Patel, A.S., "Factors Determining the Level of Fertiliser Use: A Study in Two Districts of Gujarat State", *Artha Vikas*, Vol. 7 (1), January-June 1971, pp. 22–39.

Patel, B.B., "Consumption of Fertilisers in Developing Countries: Relevance of the Price Factor", *Economic and Political Weekly*, Vol. 3 (26–28), Special Number, July 1968, pp. 1059–62.

Patel, G.A. and Shah, R.L., "Crop-wise Consumption of Fertiliser in Bulsar and Surat District of Gujarat", *Fertiliser News*, Vol. 17 (5), May 1972, pp. 47–53.

Patel, J.S., "Gaps between Agricultural Research and its Application", *Agricultural Situation in India*, Vol. 15 (5), August 1960, pp. 457–464.

Patel, S.M., "Use of Incremental Cost Benefit Ratio in Evaluating Fertiliser Recommendations", *Fertiliser News*, Vol. 20 (9) September 1975, pp. 18–21.

Planning Commission, Study on the Use of Fertilisers and Manures in Agriculture, PEO, GoI, New Delhi, 1967.

Prashad, Rajendra, "Research on Nitrification Inhibitors and Slow Release Nitrogen Fertilisers in India: Prospects for Their Use and Problems" in Fertilisers Association of India (eds.) *FAI-FAO Seminar on Optimising Agricultural Production Under Limited Availability*, FAI, New Delhi, 1975, pp. 67–85.

Raghavachari, M.V., "Growth of Budgetary Subsidies of Central Government", *Economic and Political Weekly,* March 3, 1979.

Rajagopalan, R. et al., "Changes in the Application of Fertiliser and Economics of Fertiliser Use: A Case Study", *Agricultural Situation in India*, Vol. 18 (8), November 1963, pp. 515–21.

Ramamoorthy, B. et al., "Soil Testing a Base for Efficient Fertiliser Use for Hybrid Bajra in Draught Affected Areas", *Fertiliser News*, Vol.14 (8), August 1969, pp. 25–36.

Rao, C.H.H., "Factor Endowments, Technology and Farm Employment", *Economic and Political Weekly*, Vol. 11 (39), September 25, 1976, pp. A117–A123.

Rao, K.S., "Some Economic Aspects of Fertiliser Use in India", *Agricultural Situation in India*, Vol. 15(12), March 1962, pp. 1215–16.

Rao, C.H.H. et al., *Unstable Agriculture and Droughts*, Vikas Publishers, New Delhi, 1988.

Rao, C.H.H., *Technological Change and Distribution of Gains in Indian Agriculture*, Macmillan Company of India, Delhi, 1975.

Ray, S.K. and Sharma, H.L., "Contribution of Fertilizers and Cropping Intensity to Differential Agricultural Growth in India: A District-wise Approach", *Agricultural Situation in India*, April 1985.

Reddy, V.R. and Deshpandey, R.S., "Input Subsidy: Whither the Direction of Policy Change", *Indian Journal of Agricultural Economics*, Vol. 47 (3), July-September 1992.

Rehello, N.S.P. et al., "The Impact of the Increase in the Prices of Inputs on the Profitability and Production of Sugarcane and Paddy in Mandya District of Karnataka", *Indian Journal of Agricultural Economics*, Vol. 31 (3), July-September 1976, pp. 71–81.

Sagar, V., "Fertiliser Pricing: Are Subsidies Essential", *Economic and Political Weekly*, December 14, 1991.

Sah, D.C. and Shah, A., "Interlinked Extension and Sales of Fertiliser: Role of Industry", in ed. S.P. Seetharamanan, *Agriculture Input Marketing*, Oxford and IBH Publishing Co., New Delhi, 1992.

——, "Efficiency of Fertiliser Use: Demand for Soil Testing Services in Gujarat", *Indian Journal of Agricultural Economic*, July-September 1992.

——, "Incentives for Adopting Soil Test Based Fertiliser Use: Evidences from ABC Trials", *Indian Journal of Agricultural Economic*, October-December 1992.

——, "Dryland Farming Under the Changing Resource Environment: A Case Study of Gujarat", *Arth Vijnan*, 35 (3), 1993.

——, "Farmers' Fertiliser Practices and Soil Testing Services", G.M. Desai and A. Vaidyanathan (eds.) in *Strategic Issues in Future Growth of Fertiliser Use in India*, ICAR, New Delhi and IFPRI, Washington DC, 1994.

——, "Farmers' Fertiliser Practices and Soil Testing Services: Evidences from Gujarat" G.M. Desai and A. Vaidyanathan (eds.) in, *Issues in Future Growth of Fertiliser Use in India*, ICAR, Delhi, and IFPRI, Washington, USA, 1995.

——, "Farmers' Response to Rise in Fertiliser Prices: Short Term versus Long Term Perspective", *Indian Journal of Agricultural Economics*, October–December, 1995.

——, "Inefficiency and Efficacy of Soil Testing Services: A Case Study of Gujarat", *Arth Vijnan*, 13 (1), 1995.

Sah, D.C., Fertiliser Use in Indian Agriculture: An Economic Analysis, Ph.D. Thesis, Gujarat University, Ahmedabad, (Mimeo), 1982.

——, "Subsidy Versus Price Support to Fertiliser Industry: A Theoretical Framework", *Indian Journal of Agricultural Economic*, 37 (3), July-September 1982.

——, "Bio Fertilisers in Indian Agriculture: A Case of Gujarat", A.V. Bhuleskar (ed.) in *Indian Economic Structure and Policy,* Dhaval Prakashan, Ahmedabad, 1982; Also appeared in *Economic and Political Weekly*, March 26, 1983.

——, "Livelihood Struggle: Agriculture in Tribal Gujarat", in eds., D.C. Sah and Yatindra Singh Sisodia, *Tribal Issues in India*, Rawat Publications, Jaipur, 2004.

Sengupta, K. and Banerjee, N., "Economics of Fertiliser Use", *Fertiliser News*, Vol. 15 (9), September 1970, pp. 48–51.

Shah, A. and Sah, D.C., "Promotional Marketing: Issues in Fertiliser Use", *Journal of Indian School of Political Economy*, 8(2), April-June 1996.

Shah, V.B. and Maharaja, M.H., "Growth of Fertiliser Consumption in India: Relevance of Input-Output Prices", *Fertiliser Marketing News*, Vol. 8 (11), November 1977, pp. 23–24.

Shetty, M.S., "Factor Analysis of Use of Fertilisers by Farmers", *Indian Journal of Agricultural Economics,* Vol. 24 (1), January-March 1969, pp. 50–61.

——, "Inter Farm Rates of Technological Diffusion in Indian Agriculture", *Indian Journal of Agricultural Economics,* Vol. 21 (1), January-March 1966, pp. 189–98.

Sidhu, J.S. and Sidhu, D.S., "Price-Support Versus Fertilizer Subsidy and Evaluation", *Economic and Political Weekly,* Review of Agriculture, Vol. 20 (13), March 30, 1985.

Sikder, B.P., "Central Fertiliser Pool", *Fertiliser Marketing News,* Vol. 7 (6), May 1976, pp. 1–5 and 13.

Singh, A. and Singh, N.P., "Time and Method of Fertiliser Application Under Limited Availability" in eds., Fertiliser Association of India, *FAI-FAO Seminar on Optimising Agricultural Production Under Limited Availability,* FAI, New Delhi, 1975, pp. 299–300.

Singh, D. and Goel, B.B.P.S., "Impact of Increase in the Prices of Fertilisers on the Production and Profitability of HYVs of Rice and

Wheat Cultivation", *Indian Journal of Agricultural Economics*, Vol. 31 (3), July-September 1976, pp. 105–114.

Singh, Daroga et al., "Returns from Fertilisers on Farmers' Field", *Indian Journal of Agricultural Economics*, Vol. 25 (4), October-December 1970, pp. 25–39.

Singh, Gurdev and Ramachandran, "An Approach for Demand Estimation of Fertiliser Use", *Indian Journal of Agricultural Economics,* Vol. 23 (4), October-December 1973, pp. 115–122.

Singh, I.J. et al.,"Response of Some HYV Paddy Varieties to Nitrogen: An Economic Analysis", *Indian Journal of Agricultural Economics,* Vol. 23 (3), July-September 1968, pp. 69–79.

Singh, Karam, "The Farmers' Optimum in Fertiliser Use", *Agriculture and Agro-Industries Journal,* Vol. 4 (7,8) July-August 1971, pp. 32–34.

Singh, R., "Equity in Fertiliser Subsidy Distribution", *Economic and Political Weekly,* January 17, 2004.

Singh, R.P. et al., *Study of Change in Off Take of Fertilisers in Eastern Uttar Pradesh,* Agro-Economic Research Centre, University of Allahabad, 1977.

Singh, S., Krishna, K.S. and Bapat, S.R., "Economics of Fertiliser Use Based on Experiments Conducted in Cultivators' Fields", *Fertiliser News,* Vol. 21 (3), March 1976, pp. 8–13.

Sirohi, A.S. et al., "Allocation of Nitrogenous Fertilisers among Major Foodgrain Crops in UP", *Agricultural Situation in India,* Vol. 23 (8), November 1968, pp. 825–31.

Sirohi, A.S. and Goel, S.K., "Profit and Production Maximising Nitrogen Doses for Wheat", *Fertiliser News,* Vol. 16 (8), August 1971, pp. 45–52.

Sirohi, A.S. and Sharma, B.M., "Maximisation of Returns through Farm Planning Under Fertiliser Constraints in Alipur Block of Union Territory of Delhi", *Indian Journal of Agricultural Economics,* Vol. 33 (2), April-June 1978, pp. 47–55.

Sivakumar, D., Factors Determining Fertiliser Use in South Arcote District, M.Phil Thesis, University of Madras, 1986.

Sreenivasan, N.K., "Fertiliser Production in India: An Overview", *Fertiliser News,* Vol. 21 (7), July 1976, pp. 14–20.

Srivastava, A.P., "Agricultural Production and Increasing Fertilizer Application", *Economic and Political Weekly,* March 1983.

Srivastava, D.K. and Rao, C.B., "Government Subsidies in India: Issues and Approach", Discussion Paper No. 6, National Institute for Public Finance and Policy, New Delhi, 2002.

Subbarao, K., "Incentive Policies and India's Agricultural Development", *Indian Journal of Agricultural Economics,* Vol. 40 (4), October-December 1985.

Swaminathan, M.S., "New Technology: Problems and Potentialities" in eds., C.H. Shah and C.N. Vakil, *Agricultural Development of India: Policy and Problems,* Orient Longman, 1979, pp. 451–486.

——, "Role of Fertilisers in Realising Agricultural Production Prospects in India", *Fertiliser News,* Vol. 25 (1), January 1980, pp. 17–24.

——, "Synergistic Effect of Coordinated Use of Fertiliser and Other Inputs", *Fertiliser News,* Vol. 16 (1), January 1974, pp. 45–57.

Tambad, S.B., "Economics of Fertiliser Use: Methodological Issues" *Fertiliser Marketing News,* Vol. 6 (7), July 1975, pp. 1–3.

Tandan, H.L.S., "Dynamics of Fertiliser Nitrogen in Indian Soils: Usage, Transformation and Crop Removal of N", *Fertiliser News,* Vol. 19(7), July 1974, pp. 3–9.

——, "Dynamics of Fertiliser Nitrogen in Indian Soils: Efficiency, Loss and Long Term Effects of N", *Fertiliser News,* Vol. 19 (8), August 1974, pp. 14–21.

Timmer, C. Peter, "Fertiliser and Food Policy in LDCs", *Food Policy,* February 1976, pp. 143–54.

——, "The Political Economy of Rice in Asia: Lessons and Implications", *Food Research Institute Studies,* Vol. 14 (4), 1975, pp. 419–32.

Tomar, V.S. and Ghildyal, B.P., "Management of Soil Water and Fertiliser Use", *Fertiliser News,* Vol. 22 (4), April 1977, pp. 15–17.

Upadhyay, R.M. and Sunder, Shyam, "Economics of Fertiliser Use in Soil Test Bases", *Fertiliser News,* Vol. 14 (6), June 1969, pp. 25–26.

Vaidyanathan, A., "Constraints on Growth and Policy Options", *Economic and Political Weekly,* Vol. 12 (38), September 17, 1977, pp. 1643–50.

——, "HYV and Fertilisers: Synergy or Substitution? A Comment", *Economic and Political Weekly,* Vol. 13 (25), June 24, 1978, pp. 1031–35.

——, "Performance and Prospects of Crop Production in India", *Economic and Political Weekly*, Vol. 12 (33–34), *Special Number,* 1977.

——, "Agricultural Input Subsidy", *Agricultural Situation in India,* Vol. 57 (5), August 2000, pp. 261–66.

——, "Irrigation and Agricultural Growth", Presidential Address to 47th Annual Conference of Indian Society of Agricultural Economics, North Bengal University, West Bengal, September 23, 1987.

——, "Reporteur's Reports", *Indian Journal of Agricultural Economics,* Vol. 41 (3), July-September, 1986.

Vaidyanathan, A. and Desai, G.M., "Introduction", in eds., G.M. Desai and A. Vaidyanathan, *Strategic Issues in Future Growth of Fertiliser Use in India,* ICAR, New Delhi and IFPRI, Washington DC, 1994.

Venkateshwarlu, S. and Sen, Anindya, "Fertiliser Industry in India: Moulded by Government Policy", *Economic and Political Weekly,* Vol. 37 (4), January 26, 2002, pp. 326–36.

Vyas, V.S., "The New Strategy: Lessons of First Three Years", *Economic and Political Weekly,* Vol. 3 (43), October 26, 1968, pp. A9–14.

Index